Antje Sabine Naegeli

Lebwohl, mein lieber Hund

Abschied nehmen
von einem Gefährten

DU LIEBE FREUNDIN!
DU LIEBER FREUND!

Dass du mir deine Trauer, deinen Schmerz um den Verlust deines Hundes anvertraust, ist mir Geschenk. Ich danke dir, dass ich darum wissen und dir darin nahe sein darf. Niemand ahnt wohl, wie weh dir sein Tod immer noch tut. Du weißt es gut zu verbergen. Menschen ohne Einfühlung, die jeglichen Verständnisses entbehren, gibt es ja leider mehr als genug: „Es war doch nur ein Tier …"

An Ahnungslosigkeit und Seelenblindheit ist eine derartige Aussage wohl kaum zu überbieten. Sie zeigt uns, wie tief sich viele Menschen von unseren Mitgeschöpfen entfremdet haben, dass sie den besonderen Wert eines Tieres nicht mehr kennen. Und so weißt du oft nicht, ob du erzählen sollst, was dich bewegt. Unsensible Reaktionen kannst du jetzt wahrhaftig nicht gebrauchen. Selbst wenn du auf Menschen triffst, die selbst schon einmal den Verlust ihres Hundes erlitten haben, so ist es nicht gesagt, dass sie ihre Trauer in ähnlicher Weise erlebt haben, wie es sich für Dich anfühlt. Du bist ein empfindsamer Mensch, bist zu intensiver Wahrnehmung fähig. Das gilt nicht für jeden. Leben, so wie es sich in unserer Gesellschaft gestaltet, macht viele stumpf. Jede Trauer ist anders. Und jede Mensch-Tier-Beziehung gestaltet sich anders. Aber gewiss wirst du auch auf Menschen stoßen, die eine ähnlich intensive Bindung zu ihrem Haustier hatten, die sich darin einfühlen können, was du gerade durchmachst. Ihr Mitgefühl wünsche ich dir.

Ich trage noch so viele Bilder in mir von euch beiden, sehe dich mit deinem vierbeinigen Gefährten auf den vielen Spazierwegen, auf denen ihr immer wieder unterwegs wart, habe das vorfreudige Bellen im Ohr, wenn er spürte, dass es jetzt nach draußen gehen sollte. Wie oft konnte ich beobachten, dass ihr in großer Feinheit aufeinander abgestimmt wart. Nie musstest du einen scharfen, lauten Befehl geben. Ein sachter Ruf, und er reagierte. Du hast dich mit ihm und für ihn gefreut, wenn er sich nach Hundeart austoben und seiner Lebensfreude Ausdruck geben konnte.

Wie oft haben dich seine wilden übermütigen Sprünge zum Lachen gebracht. Wie oft hat dich seine Ausgelassenheit angesteckt und jede Menge Wohlgefühl in dir geweckt. Diese reine unverfälschte Freude hat etwas so Faszinierendes, nicht wahr?

Dein Vierbeiner war ein so freundliches, zutrauliches Tier mit einer ganz eigenen „Persönlichkeit“. Er hatte ein feines Gespür für Menschen, mochte aber nicht jeden.
Wenn ich dich besuchte, wusste ich mich nicht nur von dir, sondern auch von ihm aufs Schönste willkommen geheißen, sodass mir das Herz aufging. Ich durfte dann zu seinem „Rudel“ dazugehören und habe das sehr zu schätzen gewusst. Die Erinnerung daran wärmt mich bis heute.

Freude

Wie viele Sonnenaufgänge
in meinem Herzen
wenn du mich begrüsst hast
mit wilden Sprüngen!
All deine Freude
hast du an mich verschwendet.
Welch ein Willkommen,
Bruder Hund!

Auch wenn ich weiß, dass wir Tiere nicht vermenschlichen dürfen, so wage ich dennoch zu sagen, dass ihr beide, du und dein Vierbeiner, eine besondere Beziehung zueinander hattet. Manchmal sind Hunde beziehungsfähiger als Menschen, wenn auch auf einer anderen, aber doch nicht unähnlichen Ebene. Sie sind hingabefähig, beständig in ihrer Zuwendung, empathisch, sensibel, liebevoll und kaum egoistisch. Und sie können eifersüchtig sein. Forscher haben herausgefunden, dass sie das vom Menschen her bekannte wichtige Bindungshormon Oxytozin verstärkt ausschütten, wenn sie mit ihrem menschlichen Lebensgefährten Zeit verbringen. Das ist einfach staunenswert, scheint mir. Welch ein Geschenk, einen tierischen Freund zu haben!

Welch eine Bereicherung für unser Leben.
Ich zweifle nicht daran, dass auch unsere Haustiere eine Seele haben, dass ihnen die ganze Gefühlsskala zu eigen ist, die auch zu uns Menschen gehört. Wer das nicht spürt, dem weiß ich auch nicht zu helfen. Zumindest können wir uns von der Forschung darüber belehren lassen, dass unsere Hunde ein dem Menschen vergleichbares soziales Gehirn haben, was ihre Beziehungsfä-

higkeit bestätigt. Aber eigentlich bedürfen wir als Tierfreundinnen dieser fachlichen Erkenntnis nicht. Wer sein Leben mit einem Hundegefährten teilt, ihn und seine Mitgeschöpfe mit dem Blick des Herzens wahrnimmt, kann aus Erfahrung bestätigen, dass ein Hund ein Herz auf vier Pfoten ist, vorausgesetzt natürlich, dass Menschen ihm keinen Schaden zugefügt haben, der ihn in seinem Wesen verändert und beschädigt hat. Ich muss an Fotos und Berichte denken, die immer wieder einmal in der Medienlandschaft erscheinen und mit Sicherheit keine Erfindung phantasiebegabter Autoren sind, dass nämlich Hunde wochenlang die Nahrung verweigern, wenn sie ihren Menschen verloren haben. Vor wenigen Jahren verstarb in General Roca in Argentinien der Besitzer eines Hundes. Vierbeiner Bobby weicht seither seinem auf dem Friedhof begrabenen Herrchen nicht mehr von der Seite und verbringt den Großteil jeden Tages am Grab seines ehemaligen Besitzers. Dies ist nur ein Bericht, der aber für zahlreiche andere steht, die weltweit Ähnliches zu erzählen wissen.

Längst ist erwiesen, dass da nicht einfach der fehlende Ernährer betrauert wird, denn diese Tiere sind in aller Regel versorgt, oft von Familienangehörigen. Und doch zieht das Vermissen sie auf den Friedhof. Für mich ist dieses Verhalten das Zeugnis einer großen Bindungsfähigkeit, und ich zweifle wie viele Fachleute nicht daran, dass Tiere sehr tief trauern können, weil sie beziehungsfähige Geschöpfe sind. Welche Überheblichkeit, sie lediglich zu einem Bündel aus Triebkräften zu degradieren. Wir unterschätzen unsere Tiere maßlos, wenn wir ihnen die Fähigkeit absprechen, ähnlich fühlen zu können wie wir Menschen.

Sie sind uns wesensmäßig sehr nahe.

Wie sehr ihr beide von Seele zu Seele miteinander verbunden wart, das habe ich, wenn ich mit euch zusammen sein durfte, oft gespürt. Nicht nur du hast das Ergehen deines Hundegefährten mitgefühlt und ihn fein wahrgenommen. Auch er hat deine jeweilige innere Verfasstheit spürbar ertastet und darauf reagiert, unbestechlicher und deutlicher vielleicht als mancher Mensch um dich herum. Wer nie einen Hundegefährten an seiner Seite hatte, mag das für Einbildung halten.

Ich aber habe es nicht nur an euch beiden beobachtet, sondern habe es selbst erfahren durch meinen Labrador, der manche Lebensjahre mit mir und meinen Kindern geteilt hat. Wie oft hat er behutsam, fast möchte ich sagen zärtlich den Kopf auf meinen Schoß gelegt, wenn ich traurig war, und wie oft hat er meine Fröhlichkeit mit Übermut beantwortet.
Nichts erklären zu müssen, nicht Gefahr laufen, missverstanden zu werden, bedingungslose Zuwendung zu erhalten, das ist ein Hundegeschenk, nicht wahr?

Für mich gibt es keinen Zweifel, dass unsere Haustiere eben nicht nur reflexgesteuerte Wesen sind, sondern dass sie auch lieben können. Eine bewegende Geschichte fällt mir dazu ein, von der ich dir einmal erzählte, weil ich sie so beglückend fand: Ich las von einem Pferd in Schweden, das mit seinem Bein zwischen zwei Bretter geraten war, sich dabei verletzte und sich bei dem Versuch, sich zu befreien, auch noch eine Kopfwunde zuzog. Es fiel in einen Schockzustand. Der Stress drohte das Tier kollabieren zu lassen. Sauerstoff und Kortison blieben wirkungslos. Der Tierarzt kapitulierte. Die Stute kämpfte mit dem Tod.

Da schlich sich die zum Hof gehörige Bulldogge zu ihr in die Box, kuschelte sich an das Pferd und blieb tagelang in dieser Körper- und gewiss auch Seelennähe zu ihm. Das Wunder geschah. Was menschliche Hilfe nicht zustande brachte, gelang dem Hund, der später als Lebensretter zum Hund des Jahres gewählt wurde.

Wie viel uns unsere Hunde zu geben haben, muss ich dir nicht erzählen. Du hast es über viele Jahre erfahren. Umso tiefer ist der Schmerz, dass eine Beziehung die dein Leben spürbar geprägt hat, nun verloren gegangen ist. Die Atmosphäre im Haus ist deutlich verändert. Auf welch wohltuende Weise war sie all die Jahre mitgeprägt schon durch die bloße Gegenwart deines Hundefreundes. Wie konnte er Frieden verbreiten, wenn er ruhig auf seiner Matte lag und schlief oder still beobachtete, was um ihn herum geschah. Der Platz, der deinem vierbeinigen Freund zugehörte, ist nun leer. Keine feuchte Hundeschnauze weckt dich am Morgen mit einem Stupser, kein fragender Blick erinnert dich daran, dass es Zeit ist für das Futter oder den Spaziergang. Kein aufgeregtes Bellen, das einen Besucher ankündigt, durchdringt die Stille. Futternapf, Hundespielzeug und Leine hast du noch nicht wegräumen können. Ihr Anblick schmerzt. Wenn du von unterwegs heimkehrst, gibt es nicht die stürmische Begrüssung, die dir so oft das Herz gewärmt hat. Die Matte, sein Ruheplatz ist verwaist. Wie solltest du nicht traurig sein, ist doch ein Freund von dir gegangen.

Du hast seinen Tod kommen sehen, hast mit ihm gelitten, als er immer schwächer wurde und nicht mehr wollte, was doch zum täglichen Miteinander gehörte: hinausgehen, spielen, die Umgebung mit Augen und Nase erkunden, Artgenossen beschnuppern, Leckerlis schnappen.
Es hat dich geschmerzt, dass du ihm nicht helfen konntest, obwohl du alles Menschenmögliche für ihn getan hast. Aber deine Zuwendung hat er ganz gewiss bis zuletzt in sich aufgenommen, da hege ich keinen Zweifel. Du hast bei ihm gewacht, als er die letzten Stunden auf dieser Welt durchlebte, hast deine Hände behutsam auf seinen Körper gelegt und ihm zärtliche Worte zugeflüstert, bis sein Atem versiegte.

Eine Zeitlang warst du unsicher, ob es zumutbar wäre, ihn auf natürliche Weise aus dem Leben gehen zu lassen oder ob ärztliches Eingreifen notwendig werden könnte. Dass Letzteres unterbleiben konnte, hat dich begreiflicherweise erleichtert, denn die Entscheidung über Leben und Tod deines Weggefährten wäre letztlich bei dir gelegen. Wie gut, dass dir dies erspart geblieben ist.

Denn so, wie ich dich kenne, hättest du lange darunter gelitten, über den Zeitpunkt seines Todes bestimmt zu haben, ohne unbezweifelbare Gewissheit, dass diese Entscheidung richtig gewesen wäre. Ich kenne einige Menschen, die lange daran zu tragen hatten, dass sie über Leben und Tod ihres geliebten Hundes entschieden hatten. Aber ich weiß auch, dass du dir um seinetwillen eine solche Entscheidung abverlangt hättest, wenn er sich im Übermaß hätte quälen müssen.

Nichts ist jetzt so gegenwärtig wie seine Abwesenheit. Tag um Tag bist du mit ihr konfrontiert. Dein Herz ist abschiedsschwer. Wir wissen es beide: Es gibt nichts, das diese Abwesenheit abmildern oder aufheben könnte. Auch ein anderer, neuer Hund könnte das nicht tun. Schon mehrmals haben dir Menschen aus deinem Umfeld geraten, möglichst schnell wieder ein neues Tier anzuschaffen. Aber du kannst es dir nicht vorstellen, dass ein anderes Tier den Platz deines Hundes in Haus und Herz einnehmen könnte. Es käme dir vor wie ein Verrat.

Bruder Hund

Seite an Seite
sind wir gegangen.
Haben uns so viel gesagt
von Augenpaar
zu Augenpaar.
Wussten so viel voneinander
von Seele
zu Seele.
Nie mehr
werde ich verlieren,
was du geweckt hast
in mir.
Mein Dank
geht dir nach
über die letzte Grenze hinaus,
dorthin, wo du Heimat gefunden hast.
Adieu Bruder Hund.

Deine Trauer lässt sich nicht umgehen, und das würdest du auch nicht wollen, denn du weißt sehr wohl, dass diese Trauer der Preis ist für etwas Kostbares und Schönes, das dir mit diesem Tier zuteilwerden durfte. Unfassbar ist es, dass dein Schützling nicht mehr am Leben ist. Es gibt keinen anderen Weg, als die Trauer zuzulassen, zeugt sie doch von dem großen Wert, den das Miteinander mit deinem Hund in dein Leben gebracht hat. Die gegenseitige Teilhabe an euer beider Leben war ein tägliches Geschenk, selbst wenn es zwischendrin mal Ärgerlichkeiten und Mühseligkeiten gab. Sie hatten kein Gewicht, das gezählt hätte. Wir wissen es beide: Je tiefer und inniger die Beziehung zu einem geliebten Wesen war, desto durchdringender zeigt sich der Trennungsschmerz. Den Verlust beweinen zu können ist wichtig und natürlich, denn Beziehung ist etwas vom Wertvollsten in einem Menschenleben. Auch die Beziehung zu einem Tier. Ein Wesen hergeben zu müssen, das einem zutiefst ans Herz gewachsen ist, gehört zu den Erfahrungen, die uns immer mit einer großen, unbarmherzigen Härte treffen.

In diesen Wochen stehen dir vor allem die Bilder aus den letzten Wochen mit deinem treuen Begleiter innerlich vor Augen. Das ist mehr als verständlich, denn sein Leiden und Sterben hat dich zutiefst mitgenommen, hat dich angegriffen, ist dir unter die Haut gegangen. Deine Seele braucht Zeit, um zu bewältigen, was passiert ist. Ich vermute, dass nach und nach auch wieder Bilder aus gesunden Zeiten ihren Raum in dir freigeben werden. Weißt du noch wie die ersten Wochen verliefen, als er zu dir ins Haus kam? Zum ersten Mal wurde der kleine Welpe nach gut einem Vierteljahr von seiner Hundemama getrennt.

Du hast den kleinen Kerl nach und nach an sein neues Zuhause gewöhnt, hast viel Geduld aufgewandt, bis er lernte, draußen sein Geschäft zu erledigen, auf Befehle zu hören und an der Leine zu laufen, ohne zu zerren und zu ziehen. Auch wenn dich dies alles viel Zeit und Mühe gekostet hat, weiß ich noch gut, wie beglückt du warst über den neuen Hausgenossen und wie nachsichtig du zur Kenntnis nahmst, wenn er wieder einmal Sachen zerbissen hatte mit den scharfen spitzen Zähnchen.

Schon nach kurzer Zeit konntest du dir deinen Alltag ohne deinen Hundegefährten nicht mehr vorstellen. Vielleicht magst du bei Gelegenheit wieder einmal einige Fotos mit mir anschauen, die du in dieser Zeit aufgenommen hast. Lass es mich wissen, wenn du innerlich soweit bist, dies zu tun. Was für ein kraftvoller Kerl ist aus dem kleinen Welpenbündel geworden, der nicht nur dich, sondern auch dein Haus gut bewacht hat. Und wie fremd fühlt sich dein Zuhause ohne deinen Liebling jetzt an.

Wie seid ihr zusammengewachsen in all den Jahren, in denen ihr Leben geteilt habt. Du hast mir einmal anvertraut, dass du durch das Miteinander mit Malte noch einfühlsamer und sensibler geworden bist im Hinblick auf das Ergehen der Tiere in dieser Welt, dass du mehr und tiefer leidest als zuvor, wenn es Tieren nicht gut geht, sie schlecht behandelt oder nicht artgerecht gehalten werden. Das kann ich gut mitfühlen, ist es mir doch ähnlich ergangen mit meinem Hundegefährten.

Welch ein Geschenk sind unsere Mitgeschöpfe! Wie viel Freude wecken sie in uns! Allein schon durch ihr bloßes Dasein können sie unsere seelischen und körperlichen Schmerzen lindern, können uns trösten und beleben. Wer einen tierischen Freund hat, ist weniger einsam, mag es ihm auch an menschlicher Nähe mangeln. Tiere kennen und lieben zu lernen, verändert uns in wesentlicher Weise. Sie sind uns anvertraut vom Schöpfer der Welt, unsere tierischen Freunde, sind alle miteinander unserer Fürsorge, unserer Sorgfalt und Achtung, unseres Schutzes bedürftig und wert.

Ich zweifle nicht, dass die Tiere als Mitgeschöpfe eine Aufgabe für uns haben, vielleicht die, Bindungsfähigkeit, Mitgefühl, Zartheit, Liebe und Verantwortlichkeit in uns zu wecken und uns das Staunen zu lehren über die unermessliche Vielfalt und Schönheit, der wir in ihnen begegnen. Albert Schweitzer nennt es „Ehrfurcht vor dem Leben." Wenn ich daran denke, welch ein Wunder nur schon die Augen deines Hundes waren, wie sie sein Fühlen widerspiegelten und zu reden verstanden, das hat mich immer wieder angerührt.

Dank an Franziskus

Du hast uns gelehrt
Bruder und Schwester
zu sagen
zu unseren Mitgeschöpfen,
den Tieren.
Und wie sehr
wünsche ich mir,
dass auch sie Grund finden
in uns Menschen
Brüder und Schwestern zu erkennen,
die ihre Würde zu achten wissen.

Dieser offene, unverstellte, arglose Blick deines Hundes, mit dem er uns anschaute. Wie wunderbar! „Bruder Hund“, diese liebevolle Benennung von Franz von Assisi kommt mir in den Sinn, wenn ich mich daran erinnere. Ja, dein Gefährte war in der Tat so ein „Bruder Hund“.
Wie groß kann unsere Freude sein, wenn wir staunen können über alles, was da mit uns lebt auf unserem Planeten.

Dein Weggefährte hat ein gutes Leben gehabt bei dir. Niemals war er ein willenloser Befehlsempfänger. Du hast mit viel Sorgfalt darauf geachtet, dass er in jeder Hinsicht bekam, was er brauchte. Dass du ihm ein solches Leben ermöglicht hast, mag dir vielleicht eine Genugtuung sein in dieser Zeit der Trauer. Und überhaupt: Einem Tier wohlzutun mit Zuwendung und förderlichen Lebensbedingungen tut auch uns selbst gut. Ich weiß, dass es dir immer wieder große Freude bereitet hat, deinem Vierbeiner Gutes zu tun. Auf seine Weise hat er dir viele Male dafür gedankt.

Und ich möchte dir an dieser Stelle auch dafür danken, denn miterleben zu dürfen, wie liebevoll dein Hund umsorgt war, hat auch mich im Innersten gewärmt.

Ich weiß, dass du deinem Hundegefährten gönnst, dass er nicht mehr leiden muss. In der Zeit, als es ihm gesundheitlich immer schlechter ging und du wusstest, dass es keine wirkliche Hilfe mehr für ihn geben konnte, hast du ihm manchmal gewünscht, dass die Wochen und Tage seines Leidens verkürzt werden möchten und er gehen dürfte. Dann bist du ganz in deinem Mitgefühl aufgegangen. Aber jetzt, da du ohne ihn zurückgeblieben bist, ist der Verlustschmerz überwältigend groß. Ein Miteinander tiefer Vertrautheit, das dein Leben nachhaltig bereichert hat, ist dir entrissen worden. Du hast in den letzten Jahren manchmal mit Wehmut daran gedacht, dass dein Hundefreund, wenn alles seinen normalen Gang ginge, lange vor dir sterben müsste. Aber dem Wissen darum wolltest du bewusst nicht allzu viel Raum einräumen, sondern im Heute leben und genießen, was eure gemeinsamen Tage ausmachte.

Ich denke, dass dies sinnvoll und richtig war. „Alles hat seine Zeit“, heißt es in der Bibel im Buch des Predigers Salomo. Und jetzt ist die Zeit der Trauer.

Wisse, dass der Schmerz in der Intensität, in der er sich jetzt zeigt, nicht bleiben wird. Das schwarze Loch, in das du gefallen bist, wird nach und nach schwinden, und die Tage werden heller werden. Du wirst weniger weinen als in diesen Tagen und Wochen. Noch kannst du dir das gar nicht vorstellen. Das ist völlig verständlich, aber eines Tages wirst du es erfahren. Dein Hund wird immer seinen Platz behalten in deinem Herzen. Das wissen wir beide. Und wenn du nach einiger Zeit an ihn zurückdenken wirst, werden nach den schmerzvollen auch die schönen Erinnerungen an ihn zurückkehren.

Aber auch dies wird eine Realität sein, dass du noch lange Heimweh nach ihm haben wirst. Jeder Hund, der dir begegnet, wird dich an ihn erinnern und dich spüren lassen, wie sehr er dir fehlt und wie gerne du ihn noch an deiner Seite gehabt hättest.

Ob ich glaube, dass es für deinen Hundegefährten ein Weiterleben nach dem Tod gibt, hast du mich gefragt. Was uns Menschen angeht, bezweifelst du dieses nicht, aber ob unsere Tiere auch in ein anderes Sein hinübergehen, da bist du dir sehr unsicher, bist am Tasten und Fragen.

Ich habe natürlich keine im nachweisbaren Sinn tragfähige Antwort auf deine Frage. Unsere natürliche Erkenntnis reicht ja nicht über unsere Lebenswelt hinaus. Es wäre reine Spekulation. Und doch mache ich mir meine Gedanken zu diesem Thema und will dir gerne Anteil an ihnen geben. Wenn wir Menschen, so wage ich mir vorzustellen, eingebettet sind in einen Ursprung und ein Wohin, worauf wir als Christen vertrauen, warum sollten es unsere Mitgeschöpfe dann nicht sein? Auch in ihnen schimmert die Weisheit, Zärtlichkeit und Liebe des Schöpfers, schimmert das Göttliche durch. Und wir leben ja in einer Wesensverwandtschaft und einer Schicksalsgemeinschaft mit ihnen. Wenn der verborgene Urgrund der Welt Liebe ist, können dann die Tiere aus dieser Liebe herausfallen?

Und sollte nicht auch unsere Verbundenheit mit den Tieren aufgehoben sein, ernst und wichtig genommen werden von diesem liebenden Du ebenso wie unser Trennungsschmerz? Wer allein den Menschen im Zentrum von Leben und Welt sieht, mag diese Vorstellung für sentimental oder naiv halten. Ebenso derjenige, für den mit dem leiblichen Tod alles Sein erlischt. Sei‘s drum. Ich für mein Teil trage die Hoffnung in mir, dass es auch für unsere Tiere ein Leben jenseits dieser Erde gibt.

Es gibt ja im Neuen Testament einen Hinweis darauf, dass auch die Schöpfung, also auch unsere Mitgeschöpfe, auf Erlösung wartet und sie ihr zugedacht ist.

„Auch die Schöpfung seufzt, sie ist wie eine Welt, die darauf wartet, dass etwas Neues geboren wird. Auch die Schöpfung soll von aller Verlorenheit befreit werden zur Freiheit und Herrlichkeit der Kinder Gottes.“
(Paulus im achten Kapitel des Briefes an die Gemeinde in Rom)

Zugegeben, unser Intellekt kommt da an seine Grenzen, aber wer je im Herzen Liebe, Verbundenheit, Zugehörigkeit erfahren hat, der müsste doch, so meine ich, ein Ahnen in sich tragen, dass auch unsere tierischen Geschwister heimkehren dürfen in ein Sein jenseits alles Leidens. Warum solltest du dieser Hoffnung nicht eine offene Tür gewähren?

Doch selbst wenn dies gelingt, die Trauer wird dir damit nicht erspart. Das Vermissen ist auch mit einer solchen Hoffnung nicht aus der Welt geschafft. Mag sein, es wird ein klein wenig abgemildert. Das wage ich nicht zu beurteilen.

Verweilen

Auf den Wegen mit dir
die Eile verloren.
Das Stehenbleiben
gelernt,
den ruhigen Atem,
das Schauen.
Auf den Wegen mit dir
ein Stück Leben
wiedergefunden.

Ich weiß nicht, ob du es übers Herz bringst, die Wege zu begehen, die immer wieder zu euren Hundewegen zählten. Vielleicht ist es dir ein Bedürfnis. Aber vielleicht tut es auch einfach zu sehr weh. Wenn du aber auf diese Weise unterwegs sein magst und das Bedürfnis hast, nicht allein zu gehen, so will ich dich gerne begleiten, wenn du es möchtest. Vielleicht darf ich einfach bei dir sein, ohne dass wir viel reden müssen, sondern so, dass du deinen Gedanken an deinen Gefährten nachhängen kannst.

Ich bin mit dir dankbar, dass du einen guten Ort gefunden hast, wo dein Hundefreund seinen letzten Ruheplatz gefunden hat. Es ist ein schlimmer Gedanke, unsere Haustiere nach ihrem Tod nicht würdig bestatten zu können, sondern sie wie eine Sache „entsorgen“ zu sollen. Auch unsere Tiere haben doch Achtung und Respekt verdient, und wir als ihre menschlichen Gefährten brauchen einen Ort, wo wir unsere Trauer und unseren Dank hintragen können. Es ist gut, dass du einen solchen Ort des Gedenkens hast.

Das Unverständnis mancher Menschen aus deinem Umfeld, die dem mit Stirnrunzeln begegnen, soll dich nicht verunsichern. Es gibt in diesem Bereich befremdliche Übertreibungen, das wissen wir beide. Aber von solchen Fragwürdigkeiten hast du dich immer distanziert und zwischen Mensch und Tier zu unterscheiden gewusst.

Abschied

Von der Wärme des Lebens
hast du mir erzählt,
von Leichtigkeit und Spiel,
hast mir so manchen Tag erhellt,
warst mir ein Freudenbringer,
ein Herzgefährte.

Nun geht die Trauer neben mir
und mein Schritt ist abschiedsschwer,
sagt mir, dass ein Freund
gegangen ist.
Lange noch werde ich mich umsehen
nach dir.

DU LIEBE FREUNDIN!
DU LIEBER FREUND!

wisse, dass deine Trauer mit mir geht und bei mir eine offene Tür hat. Ich fühle mit dir, was deinen Schmerz um deinen Hundegefährten angeht und ich bin von Herzen dankbar, dass er über so viele Jahre dein liebenswerter, treuer Begleiter hat sein dürfen, der dein Leben in einem ganz besonderen Maß bereichert hat, der Spuren der Freundlichkeit, der Freundschaft, der Freude auf deinem Weg zurückgelassen hat und den auch ich ins Herz geschlossen habe.

Behutsam und mitfühlend umarmt dich
und deine Trauer

DEINE ANTJE SABINE

MIX
Papier aus verantwortungsvollen Quellen
FSC® C043106
FSC
www.fsc.org

Klimaneutral
Druckprodukt
ClimatePartner.com/14549-2003-1001

Zur Autorin:
Antje Sabine Naegeli, geboren 1948 in Schleswig-Holstein, Studium der evangelischen Theologie, psychotherapeutische Ausbildung zur Logotherapeutin und Existenzanalytikerin. Lebt in St. Gallen/Schweiz und arbeitet dort in eigener Praxis. Sie ist im deutschsprachigen Raum eine gefragte Referentin. Zahlreiche Veröffentlichungen zu spirituellen und psychologischen Themen.
www.naegeli-sg.net

Zuletzt ist von Antje Sabine Naegeli im Verlag am Eschbach erschienen:
Mit mehr als nur den Augen sehen. Mit ChrisTina Naegeli (70914)
Jahresringe. Von der Lebenskunst des Älterwerdens (70869)
Du bist noch immer da (70562)
Hinter den Wolken das Licht (70768)
Entdecke, was dich ganz sein lässt (70807)

Bildnachweis:
Valentin Agapov (Umschlag), Ryan_hoel (S. 2), Paul Aniszewski (S. 6), Cosmin TARTAN (S. 11), marjan_k (S. 15), Roxana Bashyrova (S. 18), Barry Blackburn (S. 24), VarnaK (S. 30/31), Tanes Ngamsom (S. 35), alle shutterstock.

Verlagsgruppe Patmos in der Schwabenverlag AG, Ostfildern
Im Alten Rathaus/Hauptstraße 37
D-79427 Eschbach/Markgräflerland

www.verlag-am-eschbach.de

Gesamtgestaltung: Angelika Kraut, Verlag am Eschbach
Kalligrafie: Ulli Wunsch, Wehr
Herstellung: Grafisches Centrum Cuno GmbH & Co. KG, Calbe
Hergestellt in Deutschland
ISBN 978-3-86917-899-8

Manufakt
Dieser Baum steht für umweltschonende Ressourcenverwendung, individuelle Handarbeit und sorgfältige Herstellung.

PRAISE FOR

ANDROMEDA

"The subtly gorgeous prose highlights the affection between two like-minded individuals, both of whom are trying to preserve what they value the most. This is perfect for devouring in a single afternoon."

—*Publishers Weekly*

"A confident, erudite novel, comfortable with developing at its own pace... Deeply provocative in its quiet contemplation."

—*Kirkus Reviews*

"A beautifully written book about a publishing company undergoing massive changes, a questionable and yet somehow aching relationship between an intern and her boss, nostalgia, the feeling of being left behind, and the ideals we cling to when everything is falling apart around us... a breath of fresh air in the literary world."

—*Booklist*

PRAISE FOR

THE OTHER WOMAN

"Equally erotic and shrewd, the latest from Bohman (*Drowned*) reads like a confession, or diary . . . The author's prose is breathtaking, oscillating between her narrator's tumultuous feelings toward her lover and the narrator's curiosity—and occasional disdain—for the world around her . . . An elegant, rich take on an age-old narrative."

—*Publishers Weekly* (starred review)

"A woman with an almost Dostoevskian loneliness becomes the other woman in this novel of class and passion . . . Philosophical, passionate, and pensive—[*The Other Woman*] explores the psychology of both intimacy and lust."

—*Kirkus Reviews*

"*The Other Woman* hooks the reader with a captivating voice and a dramatic spin on a story we've all heard before . . . This captivating, character-driven tell-all provides the reader with a unique insight . . . And [Bohman's] bashfully charming leading lady keeps you hooked until the very last page."

—*New York Daily News*

PRAISE FOR

DROWNED

"A slim novel with a taut narrative line and a sense of impending disaster . . . A tale of identity and tense personal relationships, one that as a film property would have appealed to Hitchcock or De Palma."

—*Kirkus Reviews* (starred review)

"The seasonal and structural changes are deliberately understated and carry emotional weight into the climax, which Bohman deftly makes both foregone and suspenseful, leaving the reader wondering if everyone was doomed from the start."

—*Publishers Weekly*

"A horrifyingly compelling psychological thriller set in modern-day Sweden . . . Bohman manages to flesh out characters, setting, and plot in just over two hundred pages, a feat that makes for a highly engaging and fast-paced read, yet one with great depth."

—*Booklist*

MY STREET

ALSO BY THERESE BOHMAN

Mount Verity

Andromeda

Eventide

The Other Woman

Drowned

MY STREET

A Christmas Story

Therese Bohman

Translated from the Swedish by
Marlaine Delargy

Other Press
New York

Originally published in Swedish as *Kammakargatan: En julberättelse*
by Norstedts, Stockholm, Sweden, in 2025.

The cost of this translation was supported by a subsidy from the Swedish Arts Council, gratefully acknowledged.

SWEDISH
ARTSCOUNCIL

Production editor: Yvonne E. Cárdenas
Text designer: Patrice Sheridan
This book was set in Adobe Garamond Pro by
Alpha Design & Composition of Pittsfield, NH

1 3 5 7 9 10 8 6 4 2

 Printed in the United States of America on acid-free paper. For information write to Other Press LLC, 267 Fifth Avenue, 6th Floor, New York, NY 10016. Or visit our Web site: www.otherpress.com

Library of Congress Cataloging-in-Publication Data
Names: Bohman, Therese, 1978- author | Delargy, Marlaine translator
Title: My street : a Christmas story / Therese Bohman ; translated from the Swedish by Marlaine Delargy.
Other titles: Kammakargatan. English
Description: New York : Other Press, 2026.
Identifiers: LCCN 2026007803 (print) | LCCN 2026007804 (ebook) |
ISBN 9781635426311 hardcover | ISBN 9781635426328 ebook
Subjects: LCGFT: Christmas fiction | Novels
Classification: LCC PT9877.12.O48 K3613 2026 (print) |
LCC PT9877.12.O48 (ebook)
LC record available at https://lccn.loc.gov/2026007803
LC ebook record available at https://lccn.loc.gov/2026007804

MY STREET

HAVE YOU EVER WALKED ALONG KAMMAKAR-gatan, up the hill from Torsgatan then on down to Drottninggatan?

Have you walked along there on a Saturday in early November, when only a stone's throw away there are crowds of people out shopping or meeting friends for coffee or lunch in town, maybe looking for a Father's Day gift?

Have you noticed the striking difference between Kammakargatan and Drottninggatan, although they are so close together and actually intersect at one point?

Kammakargatan is quiet and peaceful, almost deserted, even during the weekday hours when the inner city is at its busiest. Everyone is on Drottninggatan, crisscrossing among the many shops, cafés and restaurants.

On Kammakargatan nothing happens. You go there only if you have a specific reason to do so.

For a few months in the fall of 1997, I had just such a reason.

1

THE ACCEPTANCE LETTER CAME IN JULY, I STILL remember it with an almost bizarre precision. During this period of my life I had time to spend entire mornings waiting for the mail van, and as soon as I heard it I would rush down to the mailbox. Usually I was waiting for something perfectly ordinary—a magazine, a letter, a catalog—but for the last week I had been expecting something special.

On this particular day a whole bundle of envelopes had arrived and I leafed through them frantically, found one from the Higher Education Authority and ripped it open immediately, quickly scanned the sheet of paper inside: I had been accepted onto the Cultural Studies program at the University of Stockholm, my first choice, the only thing I could imagine doing.

I sat in the shade of the birch tree on the terrace of my childhood home for a while, staring at the letter. It was strange to see my name there, in a context that had seemed so abstract for years but was about to become a reality, and I realized this meant

I was going to have to find somewhere to live in Stockholm. With a sudden burst of energy that still kicks in when I am stressed, I cycled down to the local Konsum store and bought a copy of the classified ads paper. Back beneath the birch tree I looked through the rental properties available, which didn't exactly make me feel any calmer.

Basically, I couldn't afford anything. The inner city was out of the question, as were the suburbs closest to the center. Several ads seemed too suspect to warrant a response. However, a terraced house in Skogås was advertising for a tenant, with a rent that was actually within my budget. I wasn't sure if I'd ever heard of Skogås before, I checked out the map of Stockholm in our old atlas and discovered that it was south of the city, too far out to be a part of the subway system, but there were commuter trains and I didn't have the luxury of being picky. The studio apartment had its own kitchenette and bathroom, and when I called I was informed that I was the only interested party and I was welcome to rent it.

I said yes without having seen the place, and when I did eventually see it on a cloudy August afternoon, after my parents had driven me from Kolmården to Skogås with a selection of my belongings—clothes, the compact little stereo I had received on my eighteenth birthday, my favorite records and a selection of books I felt I couldn't live without—there was no chance of backing out, because registration was due to take place the very next day. There wasn't really anything wrong with the room, except that the decor was ugly and unimaginative, there was a funny smell in the bathroom, and of course it was in Skogås.

After my parents had gone—Mom cried, which with hindsight I don't think was as over-the-top as I thought at the time, after all it was the day I left home for real—I followed my landlords' directions and walked down to the station so that I would know where I was going and how long it would take the following morning. I bought a travel card at the convenience store (dark purple, a color I have always liked) and did a little shopping in the Konsum supermarket: peppers, onions, pasta and crème fraîche, the makings of the kind of cheap pasta dish I lived on back then; bread, butter, ham, orange juice and instant coffee for breakfast; Coca-Cola, potato chips and a packet of candy for the evening—because I could make my own choices now.

I went to bed early, lay there with a book I found it hard to concentrate on, Magnus Dahlström's *Home*, which I had borrowed from the library in Norrköping (I still remember its bright yellow cover on the old pine chair that functioned as a nightstand). I was nervous about the purely practical details as usual: what if I missed the train, made a mistake when I had to change at the central station, arrived too late and didn't manage to pay my student union fee before registration, got lost around the campus? The fact that I wouldn't know anybody was also part of the picture, but it wasn't worrying in the same way—almost the reverse, in fact. It was part of the reason why I was in Stockholm: because this was going to be the start of something new.

2

EVERYTHING WENT PERFECTLY THE FOLLOWING day, and I was on campus even before the students' union had opened, almost first in line to show the little yellow receipt proving that I had paid my fee. A girl who didn't look any older than me but was clearly very adept at what she was doing gave me a blue form indicating that I was now a student. It struck me that everyone around me seemed to know exactly what to do, where to go, and above all they looked perfectly at ease, as if the fact that they were at the University of Stockholm made little difference to their lives.

It was warm and sunny, not at all the kind of idyllic autumn day I had imagined would be my first day at university, it felt like late summer, and the temperature was over twenty degrees Celsius. In order to pass the time before registration I spent a while strolling around the bookstore, Akademibokhandeln, which was also housed in the union building, but I didn't buy anything. Purchasing a hardback novel at full price was out of the question, nothing appealed to me among the paperbacks, besides which I knew that I would soon have to spend a large part of my student loan on course literature.

The foyer of the South Building was crowded, and once I had worked out the numbering system for the lecture halls it was easy to find the one I was looking for. High on the walls hung Olle Baertling's paintings featuring triangles, I had heard about them, and below them were rows of notice boards belonging to the various departments, all covered in bits of paper even though the semester hadn't yet begun. Groups of people were standing outside, weirdly a lot of them seemed to know one another already, or maybe they had just started chatting. As there were still ten minutes to go before registration I went and bought a coffee from a machine, I received a brown plastic cup containing weak but still decent coffee. I didn't have time to finish it before a woman in her fifties opened the door and everyone poured in. I was about to follow them when I saw a large notice on the door stating that the only drink allowed inside the hall was water, so I hurried over to a trash can a little way down the corridor and threw away my cup, coffee and all. The whole thing felt kind of undignified, awkward.

I went into the hall and chose a seat close to the aisle in the center, I had to get up twice for someone to pass me. Most of those present were girls, and to my disappointment they didn't really look like cultural connoisseurs. Then again, I didn't actually know what they were supposed to look like, but I had pictured them as being more special than this collection of ordinary pretty girls with backpacks and water bottles and well-organized Filofaxes marked with colorful Post-it notes.

At exactly eleven o'clock the woman who had let us in closed the door, cleared her throat and introduced herself as Margareta Holmström. She would be responsible for the first module and welcomed us to the Cultural Studies program. She quickly called the register, switched on the overhead projector and went through the outline of the program. There would be an introductory course for everyone, then freestanding modules up until a final joint semester which would also include an internship. I already knew all this and listened with half an ear, absorbed in the list of optional courses that seemed to me to sparkle like a necklace of tempting possibilities: Ancient Culture and Society. Cinematography. Art History. Literature. In just a few months' time I would have to make my choice.

Suddenly the door opened and a guy walked in. He was tall and slender, dressed entirely in black—skinny black jeans and a black shirt—with dark hair in an unspecified style. Over his shoulder he carried a bag made of worn cognac-colored leather.

Margareta looked at him.

"By the way, I forgot to mention," she said wearily. "We don't allow the academic quarter at this university."

The odd burst of laughter came from around the room. I had never heard the expression before and didn't understand what was funny, and maybe it was because I didn't laugh that the guy in black came and sat beside me. He gave me a nod, opened his bag and took out a black notebook with an oilcloth cover, and a cheap ballpoint pen.

The introductory session was soon over. Teaching wasn't due to begin until the following day, so I put away my notepad and pen, and my neighbor did the same. Before he got up to leave he turned to me and held out his hand, which was cool from the fierce air conditioning.

"Andreas," he said with casual politeness.

"Therese," I replied, much too quickly and suddenly incredibly conscious of trying not to speak with my usual Östergötland accent, my tongue kind of tied itself in knots and it sounded as if I were saying "Triss" or "Chris" and he said "Sorry?" and I repeated my name, a little more slowly this time. He nodded.

"See you later," he said. He got to his feet, left the room and he was gone.

3

ON THE WAY BACK FROM THE UNIVERSITY I GOT off the subway at Östermalmstorg, took the escalator up to Stureplan and went into Hedengrens' bookstore. I had been there only once before and loved it, and as I stood in front of the shelves of journals by the stairs on the ground floor I was absolutely certain that I still loved it. I had read or at least heard of some of the journals, but most were new to me, exotic: both Swedish and international publications that all took the topic of literature very seriously. Others were more like ambitious fanzines with varying degrees of success when it came to design. I stayed there for a while cautiously leafing through some of them, filled with a sense of being at the hub of a kind of literary interest, where young people evidently had both the time and desire to try to spread their love of books. "STEREO. The literary fanzine. Twice as good," proclaimed the cover of the most attractive publication. A classic typeface above an image of Rilke staring into the camera, his expression grave.

It cost thirty kronor, which seemed like a perfectly reasonable amount, so I decided to buy it. I had glanced at the list of course literature we had been given at registration and decided to put off that particular project until the following day. Hardly any of the titles were especially appealing, which I found disappointing.

I carried on browsing around the ground floor of Hedengrens, it was quiet and cool and pleasantly dimly lit compared to the crowds and bright sunshine on Stureplan. In the English language fiction section all the books had such attractive colors, on one shelf there were several copies of Douglas Coupland's *Polaroids from the Dead* standing side by side, it was a square, blue book with a pop art–influenced cover. I took down a copy and leafed through it with a certain reverence. On one double-page spread I found two large black and white portraits, one of Sharon Tate and one of Charles Manson. I stared at them, they were both so good-looking, beautiful even. I was familiar with their story and the whole thing was grotesque, but they were still so beautiful. Sharon Tate's face was magnetic, I could hardly take my eyes off her, and the whole book was stunning, but expensive, almost three hundred kronor.

I quickly closed it as an assistant passed by, I carefully put it back on the shelf. Next to it there was a basket of bargain books, most of them from something called *The Wordsworth Poetry Library*, large, slightly dog-eared paperbacks, all twenty-nine kronor each. I rummaged among them, as so many people had clearly done before me, and found a single copy of D. H. Lawrence's collected poems. All I knew about Lawrence was that he had written *Lady Chatterley's Lover*, but this book was quite thick, so he must have written lots of poetry too. I glanced through it and was taken by several poems, one about a kiss on a train that had a kind of swirling rhythm that stuck in my head, and I decided to buy that too. On my way out, carrying the book and the magazine in a green bag bearing Hedengrens' logo, with the sun beaming down on Stureplan, I felt somehow contented with my life as a whole, and above all with the fact that I was now a person who bought her books at Hedengrens.

4

AT HOME IN BED IN SKOGÅS THAT EVENING I leafed through *Stereo*, the literary fanzine. My room wasn't very big, and if you didn't want to sit on the uncomfortable desk chair, the only other option was the bed, and before long I would be spending almost all my time there, eating dinner while sitting cross-legged at the foot of the bed with my plate on my lap while watching MTV or some sitcom.

A long essay about Rilke ran over a number of pages, written with both gravity and affection. It focused on his connections with Sweden and his time at the castle of Borgeby in Skåne and was beautifully illustrated with small black and white vignettes. I wasn't all that interested in Rilke, in the past I had read only a few of his poems in an anthology, but it was still fascinating to realize that someone was interested in him—presumably someone about my own age. Roman Bauer was the name of both the writer and the translator of the poems, and when I looked back at the editorial details on the first page it transpired that Roman Bauer was the fanzine's editor-in-chief, as well as "student, literary jack-of-all-trades and flâneur." Other articles included an essay about Strindberg's inferno crisis, a wide-ranging piece on Nabokov, and a number of short stories and poems by young unknown writers, which all had more or less the same tone: dark, grandiose, romantic.

I switched to the D. H. Lawrence book, where the tone was also grandiose but more pessimistic. There was a great deal about death and the fact that most people were idiots, and although it was written by an ageing, bitter man, there was something about it that appealed to me.

I overslept the following morning, which was unusual for me, but I thought maybe it was due to a kind of exhaustion after so many new impressions, and being unable to relax for a second because I was afraid of doing the wrong thing, mainly getting lost on foot or catching the wrong train, because I still didn't know my way around Stockholm, or failing to modify my Östergötland accent when someone spoke to me. Being constantly on my guard took its toll.

Feeling stressed I stood in front of the mirror in the poorly lit bathroom and wearily tried to sort out my hair, sticky with a cheap apricot-scented cream in an attempt to control the worst of the frizziness.

What did I look like back then? A young, rounded face that still lacked shadows and clearly defined features. Long, curly hair that I hadn't yet started to dye a darker color, it was a boring shade of mid-brown, just like my eyebrows, everything on my face was kind of fluid, undefined. I wore jeans most of the time, my favorite was a pair from UFF on Repslagaregatan in Norrköping, the most perfect jeans I had ever owned (and with hindsight I realize probably the most perfect I will ever own), tight over the thighs and knees then flaring to the hem, presumably from the seventies. I usually paired my jeans with a simple top, I remember a long-sleeved, fitted cinnamon-brown T-shirt from H&M that I thought suited me very well, or a black polo-neck sweater, or a dark blue one in thin wool, also from a second-hand store. I wore the same pair of Dr. Martens that had seen me through the whole of my last year at high school, maybe a pea coat from the discount store Överskottsbolaget, or later in the fall when the weather grew colder, a shearling coat from the men's department of the thrift store, it was several sizes too big for me but I thought it looked suitably bohemian.

For parties I swapped my jeans for a short black corduroy skirt and black tights, perhaps a pair of knee-high platform-heeled boots in faux suede that I had bought from a sale at Din Sko in an attempt to look like Sarah Cracknell in Saint Etienne, even though I didn't remotely resemble her in any other way. Possibly a pair of plastic clip-on earrings that I had picked up second-hand, because I didn't have my ears pierced until I was almost twenty-five.

Looking back it occurs to me that I might have given a slightly eccentric impression with my confused clothing style, which was actually down to the fact that I was constantly short of money. Then again, maybe I did actually look poor and weird.

5

IN ANY CASE MY APPEARANCE DIDN'T SEEM TO have frightened Andreas away, because at recess that same morning he accompanied me to the coffee machine where we both bought ourselves a drink, then he asked me if I'd like to go outside with him for a smoke.

"I don't smoke," I said, and he looked so surprised that I quickly added "on weekdays," and he nodded understandingly.

I went with him anyway, out onto the paved area beyond the corridor, where he lit a cigarette and gazed out over the sprawling bushes and the parking lot.

"A lot of reading to do," he commented.

"Lots of books to buy," I replied.

He nodded. "We could chip in together."

This was an idea that hadn't occurred to me.

"Yes," I said. "Er—what exactly do you mean?"

He casually knocked the ash off his cigarette into an ugly aluminum ashtray on the concrete slabs.

"We could each buy half of the books, then photocopy them."

Even though I had never studied at a university before, there was one thing I knew about course literature: you were absolutely not allowed to photocopy it. In spite of this, I nodded.

"So where would we do that?"

"I'll take care of it," he said.

I nodded again. "Okay."

"Okay," he said, and stubbed out his cigarette.

After the lecture we went to Akademibokhandeln together and headed for the cultural studies section. Everything we needed for the first part of the course was there: *Cultural Studies and the Study of Popular Culture*, *The Cultured Human*, *Cultural Studies in Swedish*, a small book with a beautifully marbled green cover entitled *Cultural Policy in Practice*. We bought the whole lot and paid half each, then Andreas took the heavy bag.

"I'll get this done by Thursday," he said. I watched him disappear beyond the barriers beside the road to the subway, behind which was a large construction site where they were finishing the building that would be known as the Aula Magna, and I wondered for a second if he really would keep his promise, or if I'd stupidly let him walk away with all my books.

However, on Thursday he was waiting outside the lecture hall when I arrived, holding the same plastic bag from Akademibokhandeln, which he handed over to me. It contained the books I had paid for, along with three thick photocopies of the books he had bought.

"Wow," I said. "Cool."

He gave me a little smile. "Coffee?"

I nodded. "Coffee."

I went outside with him for another cigarette, the sky was gray, overcast, at last you could begin to sense that fall was on its way. Andreas was wearing a black jacket over his black shirt.

"We're having a little gathering on Saturday," he suddenly said. "You're welcome to come along if you want."

"That would be great," I replied, I was surprised and I had no idea who he meant by "we." Maybe he had a girlfriend, or even a boyfriend, he was hard to read, there was something faintly sensual about his face, yet at the same time something oddly asexual. Perhaps he lived in some kind of collective. So far he hadn't told me anything whatsoever about himself, nor had he asked me any questions about myself, or given the slightest impression that he wanted to hang out with me outside the university.

"Any time after eight," he added. "Feel free to bring a friend."

"So where do you live?" I asked.

"On Kammakargatan. Number twenty-seven, third floor. Ring the bell next to Hammarsten."

6

NEEDLESS TO SAY I DIDN'T HAVE A FRIEND TO take with me. I didn't know a single person in Stockholm, and my social successes after a week at the university were virtually non-existent.

So on Saturday evening I walked alone from Central Station to Kammakargatan, a street I had never even heard of. I had had to look it up on the street map in the back of my landlords' phone book. It wasn't hard to find, it cut across the top of Drottninggatan, and as soon as I turned into it I could tell that it had a unique atmosphere. It had a slight upward slope and there was nothing special about the buildings at street level, the few commercial properties appeared to accommodate offices of some kind, they were in darkness or had closed blinds. It was very quiet, the colors of the houses seemed kind of faded, there was litter on the sidewalks. It was like a back street right in the middle of the city, and I got the feeling that this could have been Stockholm back in the day, maybe in the sixties, that there might be a beer hall somewhere, or an old-fashioned tobacconist's. I would later discover that there actually was, Nille's Tobacconist's, I would go there to buy slim menthol-flavored cigarettes because I realized that I really needed to start smoking at parties.

But on this particular evening I still didn't smoke, and when I reached the top of the street and the house with the number twenty-seven, I saw that there was a church on a corner further down in the Art Nouveau style, its tower rising above the rooftops. The light glowed beautifully through the large rose window in shades of red and yellow, like an eye made up of jewels.

What a wonderful street, I thought. *It's almost magical.*

There were double wooden doors beneath an impressive arch at number twenty-seven, I rang the bell next to the name Hammarsten just as Andreas had told me to do, and after a while the intercom crackled and the lock clicked. I pushed open the door and stepped into a hallway where a lovely old chandelier spread a soft light, there was a worn carpet held in place by brass rods on the stairs, I followed the staircase past a landing with a dusty Kentia palm in a majolica pot, up to the third floor.

The nameplate on the door said Hammarsten / Bauer. I immediately began to suspect what would be confirmed when I rang the bell and Andreas flung open the door. I could hear music and the hum of conversation, and in the dimly lit hallway behind him stood another tall guy dressed in black, similar to Andreas at first glance, but with thicker hair, broader shoulders, and something more powerful about his whole appearance.

"Therese!" Andreas exclaimed with unexpected warmth and gave me a hug. "Welcome!" He stepped to one side. "This is Roman, we both live here."

Roman held out his hand. "Nice to meet you."

"Hi," I said. "So you're . . . Roman Bauer?"

He nodded.

"How funny—so you must be the person behind that fanzine . . . *Stereo*?"

He smiled delightedly. "Have you read it?"

"I bought it for the first time just a few days ago. I remember your name."

Because it's so beautiful, I almost said, but I managed to stop myself in time. Roman looked flattered anyway.

"Did you know that Andreas is the designer? I'm so lucky to have him. And the two of you . . . you study together, right?"

I nodded.

"Therese is the one I photocopied those books for," Andreas said to Roman, then he turned to me: "Roman and I got to know each other at the German School."

"We've been inseparable ever since," Roman said.

The energy between them was hard to define, they didn't come across like a couple, but at the same time they seemed to have a very close relationship, that was clear from the start. More like the bond between a couple of teenage girls than two heterosexual young men.

"Take off your coat," Andreas said, passing me a hanger, "and come on in. No, no, keep your shoes on," he added quickly when I began to unzip one of my boots.

The hallway alone was fantastic, illuminated only by a pair of old wall lamps on either side of a huge mirror. There was a heavy umbrella stand by the door, and on a chest of drawers stood a bunch of pale pink carnations in a blue and white Chinese ginger jar, next to a pile of old books.

When I followed Andreas and Roman into the living room I saw that the apartment continued in the same style. Everything was old. Not 1960s old, not the retro look that was so in at the time, with garish colors, teak furniture and string shelves, sixties sofas, plastic ashtrays, Stig Lindberg porcelain. Here, candles burned in pewter or brass candlesticks everywhere. Dark wooden bookcases crammed with books covered two of the walls, and around the generous open fireplace hung a hodgepodge of old paintings, which I would later discover were both authentic and valuable. On a bulbous baroque chest of drawers stood a globe spreading a lovely turquoise and orange glow, half of the countries on its world map no longer existed. In the center of the room two big sofas were placed opposite each other with a coffee table in between, there were piles of books on the table too. Music played quietly in the background, I didn't know what it was but it sounded like something old: the warm crackling accompanying it made me realize that it must be coming from a gramophone. Outside the tall windows I could see the lights of Kammakargatan.

7

NOT MANY GUESTS HAD ARRIVED YET, MOST were standing around in the kitchen, smoking or helping themselves to the wine and canapés laid out on the table. Andreas led the way.

"What would you like to drink?" he asked. "Red or white?"

I didn't really know what I preferred, but it felt as if red was better suited to this environment, so I opted for red.

"Okay, let's see," he said, looking among the bottles. "Burgundy?"

He held up a bottle as if he were a waiter in a restaurant and I examined the label, which was very pretty but meant nothing to me.

"I'm sure that will be fine," I said. "Thank you."

I didn't know a thing about wine in those days, but the Burgundy was delicious, it was a light and easy-drinking wine, I took a big gulp and immediately wanted more, so I took another gulp and suddenly my glass was half-empty, so I refilled it.

Andreas had disappeared back into the hallway when the doorbell rang again, I tried a canapé with what looked like liver pâté and some kind of jelly on it, but when I took a bite I realized it must be something else, maybe duck liver mousse which I had heard of but never tasted. It wasn't great, but it wasn't disgusting either, it had a pleasant fatty consistency, like fudge but with a meaty flavor.

As the guests in the kitchen seemed to be caught up in a discussion, I went back into the living room and sat down on one of the sofas. Andreas was standing in the hallway with the new arrival, a woman in a white fluffy jacket that looked as if it was made of downy feathers. With her white patent leather boots and blonde hair she resembled an ice queen, beneath the jacket I glimpsed a short dress covered in silver sequins. Her face was both friendly and haughty, her eyes were framed with perfectly applied black eyeliner, her cheekbones were high and clearly defined. She actually looked a lot like Sharon Tate in that picture I'd seen in the book at Hedengrens. I had never met such a beautiful woman in real life before. When Roman caught sight of her, he hurried over.

"I can't stay," she said, giving him a quick kiss on the cheek. "I just wanted to drop by and say hi."

"But you'll have a drink?" I heard Roman reply. "There's Champagne."

I hadn't been offered any Champagne, but now I watched as Roman opened the refrigerator and removed a bottle, took down two Champagne coupes from a cupboard, filled them and handed one to the woman. They toasted each other, he leaned in close and whispered something in her ear, she smiled. Then they looked around for somewhere to sit, and discovered that one of the sofas was unoccupied apart from me.

"Therese, this is Louise," Roman said as they came over. "We also know each other from school. Therese is studying with Andreas," he explained to Louise, who held out her hand, it was cool and silky smooth with nails beautifully painted in a shimmering, opalescent polish.

They sat down beside me, lowered their voices to talk about something that clearly wasn't intended for my ears. I reached for my wine glass, took a big gulp, put it down again and picked up a book from the top of one of the piles on the coffee table, an art history overview from the sixties, faded reproductions in four-color offset, the images flickered past as I leafed through the pages, big-eyed Biedermeier children and Habsburg chins. Beside me Louise laughed at something Roman had said, a delightful, sparkling laugh, and I thought that if you had a woman like Louise in your life you would do anything to make her laugh. And even though Roman was no doubt doing exactly that, she soon got to her feet, said that unfortunately she had to go, left her glass of Champagne half-full. Roman sank back down on the sofa, glanced at me.

"That was Louise."

"Yes, so you said."

He gave a slightly embarrassed laugh, ran his fingers distractedly through his hair.

"You're right. Of course I did. Sorry, I'm a little . . . tired. A little tired, that's what I am."

I nodded and waited for a continuation, but there wasn't one.

"Are you studying too?" I asked, I realized it was a hopelessly dull question as soon as I'd asked it, "Or are you a full-time editor-in-chief?" That made him smile.

"No, I'm afraid it's not exactly a lucrative industry. I'm studying philosophy. Why?" He spread his hands wide. "To be honest, I don't really know. I guess I find it interesting. But what will I become? What will I become through studying philosophy?"

"Educated?" I suggested, he gave a faint smile.

"It is Therese, isn't it?" I nodded. "You're nice, Therese. That's a good thing. We should all be nice."

It almost felt like an insult, *nice for heaven's sake* I thought, if you look like Louise you don't need to be nice. It's us ordinary country bumpkins who have to make do with being nice. But I smiled at him, murmured "Thank you," he rested his head on the back of the sofa and closed his eyes.

8

"TELL ME ABOUT LOUISE," I SAID TO ANDREAS when we were sitting in the corridor in the South Building on Monday, I was sharing my notes from the lecture because he'd arrived late.

"Louise . . . ," he said hesitantly. Something immediately happened to his tone of voice, it became chillier. "Roman and I were at the German School with her. She grew up in Östermalm, now she's studying at the Stockholm School of Economics. And she sings in some band, I think it's . . . what do you call it, house music."

I had to laugh at him, he sounded so out of touch when he brought up contemporary phenomena in which he clearly had no interest and therefore knew nothing about.

"Roman has been in love with her since high school," he added.

"Oh dear," I said. "So why has nothing happened between them? Or has it?"

He shook his head.

"Louise is just toying with him, she always has.

What she really wants are guys with money. Her father is . . ."

He mentioned the name of a man I recognized from both the news and the gossip columns, and the name of a company that I knew was part of the financial industry, although I didn't really know much about it at the time (I still don't, to be honest), but I realized that Louise had grown up as part of the absolute upper class, and that her boyfriends were usually recruited from a very different social stratum than the one Roman belonged to.

"So . . ." He allowed the sentence to die away.

"That's a shame," I said, mainly for the sake of saying something.

He shrugged. "Roman is such a damned romantic. Who spends five years pining for a girl? When the situation is completely hopeless? He needs to meet someone else."

"You think?"

Andreas nodded.

"Definitely."

9

A FEW DAYS LATER I WAS INVITED TO ANDREAS and Roman's place again, this time for something Andreas referred to as "a French evening." This turned out to involve onion soup au gratin with white bread and Gruyère cheese, which they seemed to have prepared together, and lots of bottles of French wine, the same kind as at the party, red Burgundy (which still takes me straight back to the apartment on Kammakargatan if I drink it today, a fleeting sensation of that slight lack of fresh air that characterized its rooms).

We ate at the round table in the kitchen, with a large brass candelabra in the middle that lit up everyone's faces in an almost ghostly way. From the gramophone in the living room came the sound of Charles Trenet singing about the azure sea.

"Therese, where do you live?" Roman asked conversationally. As soon as I walked in he had apologized for falling asleep on the sofa beside me at the party, and I had the feeling that this entire dinner was an attempt to make up for this faux pas.

He often began his sentences by addressing others with their first name, like a teacher who wants answers to his questions. I had never met anyone who did this, it was both flattering, because it made you feel special, and challenging, because you wanted to come up with a good response.

"In Skogås," I said.

An indefinable expression flitted across his face—recognition, I thought at first, then pity.

"I'm so sorry," he said.

I smiled. "It's not that bad," I assured him, even though I had no real reason to defend my smelly little room, but nor did I want these two worldly young men from the city center to feel sorry for me, or to regard me as some kind of charity project.

It struck me that Andreas had never asked me where I lived. Did he even know my last name? He might have seen it on the class list. Did he know where I was from? He might have worked out from my accent that I came from the provinces, and hadn't been interested enough to delve deeper into the matter.

"You live in Skogås and you've recently started on the Cultural Studies program," Roman summarized. "And you read the literary fanzine *Stereo*." He gave a small smile. "What else do you read?"

I reached down into my bag and took out the paperback of D. H. Lawrence's collected poems. I had started to carry it around so that I could read it on the train or subway. I had quickly established that it was difficult to read a novel on the train, because either there were constant interruptions from shouts and stops and other people, or I became so absorbed in the plot that I risked missing my station. Poetry was perfect: after each poem I could take a little break to orientate myself, then continue with the next one.

I held up the book and Roman nodded approvingly.

"Lawrence. Pretty much overlooked as a poet, wouldn't you say? The only thing people associate him with is *Lady Chatterley*."

I smiled and refrained from saying that I had done exactly the same until a week ago.

"They're not all good," I said. "But there are some poems I really like. I was thinking of translating a few."

I actually only thought of this as the words came out of my mouth, but Roman looked impressed.

"Not a great deal of his poetry is available in Swedish," I continued—it was a pure guess.

"Apart from Erik Blomberg's translations of course," Roman quickly replied, then seemed to regret his rather patronizing tone. "Although that's only a small selection," he added kindly.

I nodded. He held out his hand for the book and leafed through it.

"You're right, he's good," he said. "Some of it is pretty heavy—listen to this."

He cleared his throat.

A few are my fellow-men,
a few, only a few:
the mass are not

The mass are not my fellow-men,
I repudiate them as such.
Let them serve.

His voice deepened as he read, and his English pronunciation was excellent, much better than mine. It was pleasant to listen to him, more than pleasant, it was almost as if the words evoked something demonic in him, as if they gave his eyes a new fire, made his posture even more commanding. He almost spat out the final sentence, I pictured him as a prince at a Renaissance court, issuing orders to his subjects in a firm tone.

He put down the book.

"Maybe you'd like to publish a few translations in *Stereo*?" he suggested.

"Wow, that would be great!" I replied.

"Unfortunately there's no remuneration, as I'm sure you understand," he said. "Deadline November fifteenth. No extensions will be given."

10

"DID YOU HAVE A NICE TIME YESTERDAY?" Andreas asked when he saw me in the corridor the following day.

"Lovely—thank you. It was good to chat more with Roman, I didn't see much of him at the party."

"He enjoyed it too," Andreas replied, and I immediately wanted to ask how he had expressed himself, what Roman had said about me, but I held back.

"He seems to know a great deal about literature," I said instead, I wanted to keep the conversation going, but it was also perfectly true: Roman had immediately known that Erik Blomberg had translated Lawrence, and even though this wasn't totally obscure information, it wasn't exactly common knowledge either.

"He works at Alfa, the antiquarian bookstore," Andreas explained, "so it's his job to know about literature."

There was an acidity in his tone.

"That's where I photocopied our course books," he added.

"Aha." I had passed Alfa several times, it was a huge antiquarian bookstore located quite far up Drottninggatan, but it had always been closed when I walked by. However, it was easy to picture Roman in there, perhaps wearing an apron to avoid getting dust on his black shirt, perhaps sitting behind the counter with an open book in his hand, waiting for the bell over the door to ring.

The temperature had dropped significantly and Andreas had started turning up in a trenchcoat that must have been from the eighties, it was unfashionably oversized and made him look as if he were playing a spy in a thriller where his task was to smuggle microfilm.

"I'm actually meeting Roman at Valand this afternoon," he said. "You're welcome to come along if you want to."

I did want to. I would have gone along with whatever they suggested. I have often thought that those months with Andreas and Roman were in a way the best time of my life, and yet it was devastating that it came so early. From then on everything would seem like a disappointment in comparison: the people, the places, the parties, the conversations, the decor. Nothing new would ever really be able to measure up.

When Andreas and I walked into the patisserie known as Konditori Valand, I immediately realized why they wanted to meet there: the decor was fantastic, pure 1950s, perfectly preserved down to the smallest detail. We spent hours there buying one coffee after another, we talked about the next issue of *Stereo*, discussed the design of the cover. Andreas thought each issue should have a new design, including a new logo, but Roman objected loudly.

"How are people supposed to recognize it if you do that?"

"I think our readers can manage that. It'll be cool."

"This doesn't feel like *Stereo* at all."

Andreas took a few copies of *Kris* magazine out of his bag, and Roman rolled his eyes.

"Couldn't you find anything more dated?"

"Surely you can't call this dated? It's a classic," Andreas pointed out.

Roman picked up an issue with a sky-blue abstract painting on the cover and opened it at random.

"*The purpose of language is to be erased in the light of phenomenality*," he read in an affected voice, then gave Andreas a challenging look.

"I'm talking about being inspired by the design," Andreas said wearily. "Not by the contents."

"Our design is fine as it is," Roman insisted. "Therese, what do you think?"

They both turned to me, seeking my support in a conflict whose seriousness I found difficult to judge. It was hard to know what to think even when I had gotten to know them better: I often got the feeling that none of the differences of opinion between them ran especially deep, that their bickering was largely jargon, with a predictable element of drama that could seem wearying but didn't in fact mean a great deal to them. On the other hand I occasionally sensed from their tone of voice and facial expressions that there was something beyond the usual banter that could bubble up to the surface and make its presence felt, something that could make an apparently minor issue take on completely different proportions.

"I mean, I bought my first copy of *Stereo* because it looked so good," I said diplomatically, thinking that I was supporting Roman's point of view while flattering Andreas at the same time, and they both looked reasonably satisfied.

"We'll have to leave this for now," Roman said, stubbing out his cigarette and getting to his feet. "I have to go to work."

I started going to Valand on my own after that, either to study or to work on my D. H. Lawrence translation. To be honest that wasn't going too well, I still hadn't even made up my mind which poem I ought to tackle first. All I had done was to turn down the corners of lots of pages, and I had also purchased the same kind of notebook that Andreas and Roman used, with a black oilcloth cover with red edges.

Eventually I decided to try the poem that had caught my attention from the start, the one called "Kisses in the train," an early and for Lawrence uncharacteristically cheerful poem about being newly in love. This is how it begins in English:

I saw the midlands
Revolve through her hair
The fields of autumn
Stretching bare,
And sheep on the pasture
Tossed back in a scare.

I wrote, I crossed out what I'd written, wrote again, weighed the words against one another, and at last I had a first draft, the first piece of text I had translated:

Jag såg landskap passera
Genom hennes hår
Höstliga åkrar
Som upprivna sår,
En flock betande, bräkande
Vettskrämda får.

I looked at the short verse, both satisfied and dissatisfied. It was a little awkward, and I thought *nyplöjda åkrar*—newly plowed fields—might sound better, but then I would lose the reference to autumn. Maybe the wounds—*sår*—should be newly plowed? That would bring a brutal, violent element into the poem that wasn't there in the original, the use of the word *sår* did that anyway, but I didn't know how to make the rhyme work otherwise, and maybe this was what exercising poetic freedom was all about. I couldn't make up my mind, translating was more difficult than I'd expected, and I thought I would like to ask Roman, because I knew he had translated not only Rilke but also Stefan George and Hugo von Hofmannsthal and a number of other German-speaking poets I hadn't heard of before he told me about them, and I made a mental note of their names and went to the library and read their poetry. (To this day I can't get through the month of November without thinking of Stefan George's poem about a park where the plants are dying before the onset of winter, where a line in Roman's translation read "and do not forget the last pale asters.")

But I wanted to do this on my own, I wanted to prove to him that I could do it by myself.

11

THE EVENINGS AT KAMMAKARGATAN FOLlowed one another during the ensuing weeks. There was a party almost every Saturday, and as I no longer felt so ill at ease in Andreas and Roman's home, on one occasion I got into conversation with a guy called Max, who was at Konstfack—the University of Arts, Crafts and Design—and had produced illustrations for several issues of *Stereo*. His technique involved cutting up old photographs he had bought in second-hand stores, or pictures from albums he had found in dumpsters, and building "new worlds out of them," which made me ask if he didn't feel bad chopping up people's childhood memories and images from occasions that had been important to them. Maybe he was destroying the only evidence that certain things had happened, I said to him in quite an accusatory tone, and we got into a lengthy discussion about art and morality. Roman happily joined in and insisted on the importance of art being completely immoral.

I had drunk a great deal and suddenly realized that I was going to miss the last train home, but Andreas assured me that it didn't matter, I was welcome to sleep on the sofa.

And so I got to have breakfast with him and Roman too, they were up and about by nine o'clock the next day making scrambled eggs and toast, setting out marmalade and freshly squeezed orange juice and serving me coffee from an impressive Bialetti coffee maker.

Little by little I got to know the circle with which they surrounded themselves, several of their friends also lived on Kammakargatan. Max rented an apartment in the building next door, and David, an old friend of Andreas's, lived at the intersection with Västmannagatan. He played in a band called Intermission, a goth rock band who'd enjoyed modest success in the late 1980s. He was both older than us, and also the only one who was working rather than studying. He had a job at the *Dagens Nyheter* printing plant in Akalla and would sometimes turn up late to the parties with the latest edition of the newspaper still warm and sticky with printing ink. We liked to read the culture section together, with Roman and Andreas taking it in turns to pour scorn on the writers that they considered inadequate, especially those who were their own age or slightly older, writers with whom they felt some kind of rivalry.

"What we do in *Stereo* is so much better than this!" Andreas exclaimed after reading aloud an incomprehensibly glowing review of a debut short story collection.

"Twice as good!" Roman agreed, and we all laughed.

It was an exaggeration, even I knew that, however young and starry-eyed and easily impressed I still was. The literary fanzine *Stereo* was ambitious and passionate, but most of the articles were youthfully high-strung and often quite mediocre, and of the aspiring poets and authors who were published there, none would become a well-known name in the cultural world of the future (apart from me, I can probably be considered well-established in the field of literature by now, but I am also the only one who continued in that area. As time went by others became very successful, famous even, within different sectors, but that's another story altogether).

The first time I met David I was a little disappointed when he played what to my ears was impossible-to-listen-to and above all pretty embarrassing goth rock, rather than pop music. I considered myself indie and would have liked to get to know someone in Stockholm who shared my taste in music. But none of those who showed up at Kammakargatan seemed to fall into that category, none of the guys Andreas and Roman knew sported the fashionable hairstyle I thought was so incredibly cool at the time, or a tight T-shirt with a band logo, or a green parka. In terms of taste, I quickly realized that Kammakargatan was a very long way from Skånegatan and Hanna's Cellar or Tranans Bar. The clientele at Kammakargatan was interesting but not hip, it was fair to say that a fragment of 1980s elegance had survived there, a penchant for art and literature rather than pop music. Once when I suggested that we should maybe go to the Söder district for a beer, Andreas pulled a face.

"You know, the people who hang out there are just a bunch of hicks," he said. "They come from Skövde and Borås and Linköping and Motala, they wear stripy sweaters and like one another's bands and think they're so fucking cool because they know someone who writes for a magazine."

He suddenly fell silent when he remembered that I too was a hick. He cleared his throat.

"What I mean is, it's not just that they're not from the city, it's their... incredible lack of originality."

I realized that the circle who gathered at Kammakargatan consisted mainly of Stockholmers, quite snobbish individuals, elitist even, and at the same time strangely detached from the world. No one discussed politics, no one had read Nina Björk's *Under the Pink Duvet*, no one read *Pop* or *Darling* magazine, no one read Douglas Coupland, Per Hagman or Linda Skugge, or the Leisure Guide, SvD City or the Friday supplements of the evening newspapers. They didn't go to any trendy bars or cafés, instead they mostly went for coffee to the classic Stockholm patisseries, and they seemed to have no interest whatsoever in the wave of Italian food culture that had gained a foothold in Stockholm, with the shining exception of the Bialetti coffee maker. Both Andreas and Roman agreed that it was impossible to make drinkable coffee at home without it. The coffee it produced was strong with a slightly burnt taste, which I imagined was continental, even though at the time I had barely traveled overseas. Andreas told me they had bought it in Milan when they were backpacking the previous summer after graduating from high school. They were both two years older than me but seemed to have a ten-year advantage when it came to their worldly attitude, which they had turned into a blasé approach to trends and contemporary popular culture. They didn't have a TV in the apartment, and while I listened to Sarah

Cracknell's solo album *Lipslide* or The Verve's *Urban Hymns*, and Cornershop's *Brimful of Asha* was played wherever you went, the music at their parties consisted of Miles Davis, Charlie Parker and Frank Sinatra.

12

AT THE SAME TIME, ANDREAS AND I WERE TRYing to get through the first semester of the Cultural Studies program with decent grades. He missed more lectures than me, because he was often late, he would arrive after the first recess looking as if he'd just woken up, or he wouldn't come at all, there would be some evasive explanation that he'd had to take care of something. Personally I never had anything that needed taking care of and I was there on time every single morning, but I wasn't as good at sorting information as he was. Because of my unfamiliarity with the lecture format I would take notes until my pen glowed, incapable of deciding what was actually important, of distinguishing random facts from the context as a whole and major lines of development. Andreas also had a head start of several years on me when it came to concepts and people that I had never heard of back then; he could explain to me what semiotics was, he gently corrected me when I mispronounced Jacques Lacan's surname. We got into the habit of meeting up at

Valand to compare our notes so that neither of us missed out on anything vital.

"There was something I wanted to ask you," he said one day as we sat there with the last drops of cold coffee from the refill in our cups so we wouldn't have to pay for another.

He cleared his throat.

"Roman and I have been talking, and we're wondering if you'd be interested in renting the attic?"

"The attic?"

He knocked the ash off his cigarette into the already full ashtray on the table.

"My aunt owns the property, of course," he said as if he really thought I already knew that, even though no one had told me. "Roman and I live in her old apartment—she married a count and moved to Skåne."

He spoke in such a casual tone that I burst out laughing and saw a slightly irritated frown suddenly appear on his forehead.

"Sorry, it just sounded... funny," I mumbled, and he nodded.

"She has neither the desire nor the opportunity to manage the property these days, so over the last few years I've organized a few simple tasks for her. Contact with the caretaker, one or two issues with tenants. The latest tenant moved out of the attic in the spring, and it's been empty ever since. There was no urgency to let it, so I haven't made any effort to do so, but then I thought... actually it was Roman's idea, but I agree with him—it must be pretty tedious for you catching the train every day, and if you want to help out with *Stereo* too... well, we thought you might enjoy living on Kammakargatan."

It was an incredible offer, the kind I would never have dared to dream of, because such offers are usually made to people in completely different circumstances from mine. I hardly knew what to say.

"Are you serious?" I managed eventually, he looked almost offended, and for a second I was afraid that he would withdraw his suggestion.

"Of course I'm serious. You're welcome to take a look at the room. Maybe this afternoon?"

The walls were painted white and the room was sparsely furnished with an old bed, a desk and a low bookcase made of dark wood. It was actually smaller than my room in Skogås, but it was wonderful, a little apartment of my own on the top floor. The windows overlooked the courtyard, facing toward the inner city, where the neighborhood secretly revealed its rounded stairwells, and where the woodbine still glowed dark red. There was a carpet beating rack and several bicycle stands in the courtyard, a cat strolled nonchalantly across the roof of what seemed to be a toolshed. This was the heart of the very heart of the city, the most private area, protected from passersby with gates and entry codes, and now it was mine, now I was one of those who had access to it.

The shower and toilet were housed in a small cubicle, another cubicle served as a wardrobe with some old clothes hangers and a shelf covered in floral lining paper. The kitchenette consisted of two hotplates, a refrigerator and a sink, a little table with two chairs, and a few shelves on the wall with a random collection of mismatched crockery and glasses, along with a Bialetti coffee maker of course. A faded postcard with a still life by Morandi was pinned to the wall beside it.

"It's a bit unusual to have a room like this in Stockholm," Andreas said. "In France most old buildings have a *chambre de bonne*, but here it was more common to have a maid's room in the apartment itself."

I nodded as if I understood what he was talking about.

Roman had appeared in the doorway, looking delighted.

"The person who used to live here was an Estonian artist," he informed me. "He used to sit in the window drawing and listening to Arvo Pärt."

"He gave us one of his pieces," Andreas chipped in, "some kind of icon, it's hanging in the hallway,

I don't know if you've noticed it? Apparently he was quite famous. But also a drunk, unfortunately. The laundry is in the cellar, the rent is due on the last day of each month, shall we say 1,500 kronor, if you think you can afford that?"

On top of everything else, it was cheaper than the room in Skogås.

"What do you say?" Roman asked.

"I say yes, of course. Thank you so much—it's fantastic!"

They both smiled, and I remember them chorusing "Welcome to Kammakargatan!" almost as if they'd rehearsed it. (I can still see them standing in front of me, silhouetted against the light from the window, two dark figures with the light streaming through their hair, an image that in my memory has taken on almost religious proportions. But as with many things that happened during my time on Kammakargatan, I sometimes wonder if I really remember correctly.)

13

I MOVED IN THE FOLLOWING SATURDAY. I WAS supposed to give a month's notice on my room in Skogås, but Andreas had said that he would let me off the first few weeks' rent, so I decided to move right away. He also had the use of his aunt's car, which was kept in a garage close to where Kammakargatan meets Torsgatan. It was an old Mercedes, not at all suitable for moving, but then I didn't have much stuff.

We drove over to Nynäshamn on one of the last fine autumn days, it was sunny and the few remaining leaves on the trees glowed red and yellow against a clear blue sky. Andreas had tuned the car radio to P2 and the sound of swirling strings filled the air. Just before we turned off for Skogås he nodded toward a sign that told us how far it was to Brandbergen.

"That's where Roman's from," he said.

"What? Isn't he from . . . Didn't he grow up in the city?"

Andreas shook his head.

"He commuted to school every day, right up until the last year at high school, when my aunt relocated to Skåne and we were able to move into the apartment."

"Do his parents still live here?"

"Only his mother. His father lives in Germany, he went back there when they divorced. I think Roman was only three or four at the time, he has hardly any memories of his father. He doesn't usually like to talk about it, by the way."

It was very strange. Worldly Roman Bauer had grown up with a single mom in Brandbergen. I could have sworn he was just as much a city kid as Andreas, because that was how he came across: his self-assured manner, his cultured demeanor. Maybe, I thought, he was the one who had persuaded Andreas to let me move into the attic, because he thought that I too should have the opportunity to live in a place I couldn't really afford, and didn't have the connections to get anywhere near. Because he thought that I too deserved to spend some time on Kammakargatan.

Andreas couldn't quite hide his feelings about my room, even though he naturally did his best to keep up appearances. He was the perfect gentleman, he carried my two boxes of books and records and put them in the trunk. I wrapped the stereo in a blanket and placed it on the back seat, then I wheeled out the big suitcase containing all my clothes and shoes. That was everything. I shook hands with my landlords, thanked them for my time in their house, then I got into the car with Andreas and we left Skogås behind us.

14

SOON I BEGAN TO COME AND GO IN ANDREAS and Roman's apartment as if I lived there. On several evenings a week we ate dinner together, they always did the cooking because they had higher standards than me when it came to food, and I didn't want to subject them to my pasta with onions and crème fraîche. Andreas liked to prepare meatballs with mashed potatoes and cream sauce, it was his grandmother's recipe and he had mastered it to perfection, the sauce was silky smooth, its sweetness a perfect complement to the tartness of the lingonberry conserve and the slight acrid vinegary taste of the pressed cucumber salad which he served as an accompaniment. Roman occasionally wanted to eat something German, mysterious pale sausages with Sauerkraut and sweetish mustard, that was one of the few times when they bought beer, which they otherwise seemed to agree was best drunk outdoors in the summer, or if you actually happened to be in Germany.

They rarely ate out, which I didn't find strange at all, because we were students and in any case I wasn't used to going to restaurants, but one Thursday evening Andreas announced that he would like to invite us to Tennstopet. We walked there together, it was a dark evening and the sidewalks shimmered like a Byzantine mosaic made up of all the fallen autumn leaves.

Inside the restaurant I quickly realized that Andreas was a regular. The staff greeted him warmly, exchanged a few words, asked after his father and sent their best wishes. When the menus arrived and I happened to mention that I had never tried steak tartare, he insisted that we should all have it, and I couldn't really say no as he was paying. At first I thought the plate looked grotesque with its large puck of raw meat, but the taste took me by surprise: it was clean and metallic, not at all fatty and heavy as I had imagined.

"I've been coming here ever since I was little," Andreas said as we were drinking coffee after the meal. "We always celebrate my dad's birthdays here. You know how it is—when you find a place where you feel at home, you don't want to change."

It sounded so obvious when he said it that I accepted it as a truth, just as I drank in everything else they said, but then I noticed Roman trying to catch my eye. When I met his gaze he gave an almost imperceptible smile, as if he were distancing himself from Andreas. I smiled back, equally discreetly, although I was kind of embarrassed because it was generous of Andreas to treat us to dinner. At the same time I was relieved to experience a moment of mutual understanding.

In the middle of November I finally finished my translations and handed them over to Roman. I had printed them out at the university library, and thought they looked very professional even though there were only three poems, two of which were very short. Roman read them in silence, concentrating hard, then he looked up at me and nodded.

"Not bad at all," he said, which I took as warm praise.

A few days later we drank a toast to the fact that the next issue of *Stereo* had gone to press, and after another week or so we raised our glasses again when we received the printed copies, which looked fantastic. Andreas had pushed through a slight update to the design, but left the logo unchanged, so the cover looked both modern and classic at the same time. I could hardly stop looking at my translations. That was the very first time I saw my name in print. But best of all was the editorial box, where I felt that our three names stood out among the other contributors, glowing as if they were in neon: Roman Bauer. Andreas Hammarsten. Therese Bohman.

15

IT WAS DECEMBER ON KAMMAKARGATAN. THE first light dusting of snow settled and remained as the temperature dropped, brightening both the street and the courtyard and giving me an idyllic view.

One evening when I got home after spending hours in the university library studying for an exam in scientific theory, I saw Christmas decorations draped between the buildings on Kammakargatan, old-fashioned garlands, each with a shining yellow star in the center. It looked magical from the top of the hill, and the stars seemed to flow along the street in both directions.

"What lovely stars!" I said to Andreas when I met him in the stairwell.

"Beautiful," he agreed. "They're from the 1950s. The Kammakargatan Merchants' Association puts them up every year."

I laughed. "Merchants' Association? There's hardly a single store on the whole street."

He shrugged. "I think Nille organizes it, you'll have to ask him." (But I never did ask Nille, and like so many other things on Kammakargatan, its Merchants' Association would remain a mystery to me long after I moved away.)

On another evening Andreas asked me to go with him to buy a Christmas tree. We headed along Dalagatan toward Vasa Park, where children were skating. The ice was surrounded by glowing yellow lanterns, and feather-light snowflakes drifted down from the dark sky. Andreas chose a large, dense fir and we carried it home together, it smelled of forest and resin and the stars on Kammakargatan shone like a winter galaxy above our heads.

The tree was to take pride of place in the stairwell, and we managed to get it upright in the heavy old cast-iron tree stand. Then we arranged the lights, two boxes from Luma—it took us quite a while to untangle the wires.

"Thanks for your help," he said when we were finally satisfied with the result.

"It was the least I could do," I replied. "Now that I get to live here."

"We love having you here," he said in a tone that was hard to interpret, it was politely formal yet more at the same time, as if there were something between the words that he didn't want to say.

16

ONE SUNDAY I WENT WITH ROMAN AND ANdreas to the flea market on Hötorget. It opened at ten, but they were very clear that there was no point in showing up after eight if you wanted anything of value, or even anything that looked good, so we set off with just a few hours' sleep in our bodies after Saturday night's party in their apartment. It was still dark outside, there was a thin layer of snow on the streets and our breath came out of our mouths like smoke in the cold air.

The sellers were already busy unpacking their wares from removal crates and banana boxes. Roman and Andreas wandered around inspecting the items, exchanging a few words with the vendors, picking things up and examining them with a connoisseur's eye before putting them down again.

Andreas soon bought four sets of French bistro cutlery and twelve thin wine glasses with small stars etched upon them, while Roman bought a large spherical pewter vase and a pair of cufflinks in the shape of the French fleur de lys.

"Can't you find anything?" Andreas asked me.

I knew nothing whatsoever about antiques back then, plus I was pretty hungover, the only stall where I felt at home was that of an elderly man selling books and posters. I picked out a paperback copy of Marguerite Duras's *The Lover*, a book I had seen many times but not yet read.

Roman shook his head. "We've got a thousand copies of that at work, you can have one from me. You ought to buy something to hang on the wall in your room."

He led the way to a stall where there were pictures in boxes on the ground and started to look through them, checking out the signatures, I crouched down beside him. Eventually he held up a small oil painting of three white peonies, or maybe they were anemones.

"What about this?"

"It's lovely," I said.

"It's a Lotte Laserstein," he said quietly so that the seller wouldn't hear. "It will be valuable one day."

He straightened up. "What do you want for this?"

"Three hundred," replied the vendor.

Roman laughed. "Three hundred? The frame is flaking—look." He pointed to a few barely visible marks on the underside. "I'll give you a hundred."

"Okay, two hundred," said the vendor.

Roman shook his head. "We'll take it off your hands for a hundred and fifty."

The vendor shrugged, muttered "Oh, all right then" and I rummaged in my wallet, found a hundred-kronor note and a fifty-kronor note and handed them over to him. He took them without a word.

Roman tucked the painting under his arm. A dark shadow of stubble covered his cheeks and chin, and I felt as if I had accidentally seen a very private side of him, and I realized it gave me something of a thrill.

"There you go," he said as we walked away. "This is the beginning of your art collection." He made it sound obvious and ceremonial at the same time.

Andreas showed up with a stuffed pheasant tucked in among the wine glasses, which he was carrying in a banana box.

"Good hunting, I see," Roman said. Andreas smiled. "Therese has bought a Lotte Laserstein," Roman went on. "We need to celebrate—I'll treat you both to breakfast at Vetekatten."

I borrowed a hammer and nails from Andreas and hung the painting on the wall in my little room. It looked really good, the picture was bright with a pleasant color palette that fit perfectly with the almost monastic atmosphere, while adding color and life and the sense that a cultured person lived here. I gazed at the painting when I lay in bed that night, at the three peonies, or maybe they were anemones, against a pale green background, a shade I have always liked. The three flowers formed an asymmetric triangle, where two of them seemed to be supporting the third, which was turning toward the taller one of the other two. An uneven trinity.

Later I would wonder why Roman hadn't bought the painting for himself, if he knew it was valuable, or at least would be one day (he was absolutely right about that—among the quite modest but slowly growing art collection I own today, of which Laserstein's painting was indeed the beginning, it is without doubt the most valuable piece. However, as I said, there is still a great deal I wonder about when it comes to my time on Kammakargatan).

17

FOR REASONS I DON'T WANT TO GO INTO HERE, out of consideration for certain people close to me, it wasn't possible for me to go home to celebrate Christmas with my family that year. This became clear around the beginning of December, and I was surprised at how little it bothered me. If I had still been living in the room in Skogås it would have been a completely different matter: depressing on a level I didn't even want to imagine. Even today I can still feel the angst creeping up on me at the very thought. But now I was living in the attic on Kammakargatan, and even a Christmas spent alone seemed bearable there.

When the snow came the light in the room became even whiter and more monastic, and I saw the solitary Christmas holiday before me in a romantic glow. I would use the small Christmas gift my parents had deposited in my bank account to visit the Östermalm Food Hall and buy some festive food, a minimalistic Christmas table with the things I really liked, some good herring, a few slices of smoked and cured salmon, a small portion of Jansson's temptation, and a good stock of wine: red and white, Champagne if I was feeling extravagant, otherwise a cheaper but decent sparkling wine, maybe some spirits or liqueurs, hyacinths and amaryllis would bloom and fill my room with their perfume and beauty, I wouldn't need to go out for several days if I didn't want to, I could study for the last exam of the semester between New Year and Twelfth Night, until then I would just lie on my bed and watch the snow falling outside, and I would read the two green volumes of Thomas Mann's *The Magic Mountain* that I had bought from Roman at the Alfa antiquarian bookstore. He had actually been wearing an apron over his black shirt and had smiled in a both blasé and friendly way at my delight over the well-stocked shelves. Then an elderly man had approached him with complicated questions about something in the anthropology department, Roman disappeared with him for a while and came back in a much worse mood.

"This place is incredible," I said. "It must be fantastic working among all these books."

He shrugged.

"Have you always wanted to work in an antiquarian bookstore?" I continued the conversation.

"Absolutely," he said without a trace of enthusiasm. "Even as a child I sat and priced up the Swedish Tourist Association's yearbook."

I smiled at him, he smiled back, suddenly in a better mood.

"Come with me," he said, heading for the counter where the cash register was. He produced several books that he had set aside for me. Two were *The Magic Mountain*, he waved one volume in front of me.

"Do you know who translated this?" he asked, and when I shook my head he opened the front cover and showed me Karin Boye's name.

"I had no idea," I said.

"By the way, let me know if you need any help with the French section at the end," he added. "I have a translation you can borrow. And then there's this"—he held up Marguerite Duras's *The Lover*—"which I'm including for free because we must have at least twenty copies. I'm giving you a good price on *The Magic Mountain* too."

Indeed he did, and I pictured myself reading it in bed, with my window slightly ajar and the fresh winter air swirling into the room as I lit one of the slim menthol cigarettes from Nille's Tobacconist's.

And I saw myself going out one afternoon, taking a solitary stroll in a deserted Stockholm, decorated for Christmas, I might slip into a church, attend a mass or a concert, perhaps stop off in a bar and have a drink by myself, take *The Magic Mountain* out of my purse and sit there reading, watching as dusk fell outside the windows.

None of this seemed sad or embarrassing, I had a good reason for being alone at Christmas, and the scenarios I painted in my head felt nothing but pleasant.

I hadn't said anything about Christmas to Andreas or Roman, and they hadn't told me anything about their plans either. I assumed they would be celebrating with their families, Andreas in the apartment on Karlbergsvägen where he had grown up, Roman possibly in Brandbergen with his mother. I knew there was an aunt and some cousins in Stockholm too, maybe they all got together, gathered the little family in one place to celebrate in some German way, whatever that might involve. I had no knowledge of German Christmas traditions, but I pictured pretzels and marzipan, and at the same time there was a strange aesthetic dissonance in the idea of someone like Roman spending Christmas in a Million-Programme apartment in Brandbergen.

18

CHRISTMAS DECORATIONS HAD BEEN GOING UP in Andreas and Roman's apartment since the Friday before the first Sunday of Advent, but the preparations intensified as Christmas drew closer. First there were the Advent candle bridges in the living room window, old and in the Art Deco style, black and brass with probably a very unreliable electrical connection, a glowing orange Advent star in the kitchen, hyacinths and amaryllis planted in blue and white Chinese jardinières, a substantial wreath made of fir branches adorned with deep red velvet ribbons on the front door. Then came more, box after box, decorations that had belonged to Andreas's grandmother, passed on to his aunt and now to him. Two large plaster angels, both clutching a candle in each hand, were to be placed on the mantelpiece above the open fireplace, along with a thick fir garland that was always purchased from a florist up on Odenplan, the entire apartment smelled of forest when we had put it up.

Another box was full of old paper decorations, garlands made of crepe paper and shiny metallic foil, rows of paper angels and elves that would sit on bookshelves and chests of drawers. And then there were tiny porcelain holders for the Christmas tree candles, pigs and more elves and even more angels. I had never seen so many beautiful things outside the antique stores in Vasastan.

We wandered around the apartment one evening, each of us holding a glass of Champagne, and placed every item where we thought it looked best. Roman had put on an old recording of Bach's Christmas Oratorio, and a fire was crackling away in the hearth. We had found an old nativity scene with an unusually large number of figures, Roman and I carefully arranged them on a small side table next to one of the sofas, I particularly liked two rather worn palm trees. When Andreas saw what we had done, he shook his head.

"The east," he said, pointing to the wall where the fireplace was. "The three wise men come from the east. They haven't sailed to the Holy Land from Greece."

"Maybe they just took a detour to look around the city?" I said, he smiled but shook his head again and quickly rearranged the crib until it was geographically correct.

"If we're going to be picky," Roman said, lounging on the sofa with his glass of Champagne, "they didn't actually arrive until Twelfth Night."

They often teased each other, affectionately I'd always thought, but now something about the energy in the room changed, something about Andreas's slightly stiff smile made me realize that he didn't find the comment particularly funny, and that there was a reason for this that I knew nothing about.

19

EVENTUALLY THERE WERE DECORATIONS EVERYwhere that decorations could possibly go, although there were still plenty left in the boxes. Two of them contained only tree decorations, I had peeped inside and seen those elongated, pointed baubles that used to be popular, in deep jewel colors that also used to be the norm: cerise, turquoise, deep blue, and baubles shaped like pinecones and lanterns, and little glass birds to clip onto the branches.

"When do you usually trim the tree?" I asked.

"We normally wait until just before the party," Roman replied. "You are coming to the party, aren't you?"

"What party?"

There had been a party virtually every Saturday since I got to know Andreas and Roman, with varying degrees of ambition, but still some kind of party.

"Haven't you told Therese about the party?" Roman said to Andreas in an accusatory tone.

"Haven't I?" Andreas glanced at me, I shook my head.

"Tell me about the party, Andreas," I said, deliberately sounding assertive.

He reached into his inside pocket and took out a small piece of paper. It was a flyer, designed to look like a title in Bonniers' former Panache series, with a dark red frame and the little silhouette of a Greek helmet in the center at the top.

CHRISTMAS PARTY
AT
KAMMAKARGATAN

it said where the title should have been, and further down: "As usual we are celebrating Christmas at Kammakargatan 27 on December 21 from 7 p.m. Ring the bell by the name Hammarsten. Drinks and canapés will be served. Please dress up."

"We always have a Christmas party," Andreas said.

"So I see."

"You're welcome, of course."

There was something odd about his tone again. Didn't he want me to come? Had he grown tired of me, did he regret asking me to move in? Did he wish he'd found a more suitable tenant for the attic?

"I'd love to come," I said.

Roman smiled. "Of course Therese is coming to the Christmas party."

It soon became clear to me that the party, required a great deal of preparation. Apart from the decorations and everything that had to be sorted out with regard to food and Christmas treats, wine had to be purchased. This was started well in advance, because it had to be carried home in several trips. Every day the collection of bottles in a corner of the kitchen grew. I guessed that it was mainly Andreas who had paid for the wine, and I wondered what the arrangement was between him and Roman when it came to finances. Roman was a full-time student and also worked in the antiquarian bookshop. He presumably used quite a lot of his wages to pay for the printing of *Stereo*, which was nowhere near in a position to fund itself. Andreas had his studies, plus the tasks he carried out in the property on behalf of his aunt, maybe she paid him, but I suspected that he had a buffer, a savings account with money that had accumulated over time and was perhaps maintained by his parents, who were obviously well off. And since the building and the apartment belonged to his aunt, presumably he lived rent-free.

Did Roman pay rent? Or did they just split the expenditure on food and electricity and practical household items? I thought it was a big favor, allowing someone free accommodation in central Stockholm. A favor that required something in return? Explicit or implicit. Conscious or unconscious.

20

THE CLOSER WE GOT TO CHRISTMAS, THE greater the contrast between Drottninggatan and Kammakargatan. Drottninggatan bore an increasing resemblance to an inferno, swarming with stressed people dashing from store to store on the hunt for Christmas presents. You could hear the relentless Christmas music out in the street, in the entrance to Åhléns City there was a huge Christmas tree made of sage green gift boxes by Clinique, and a woman in a white overall pressed two small sachets containing samples of a new eye cream into my hand as I passed by. I only wanted to buy a pair of tights for the party, but there were so many customers waiting in line that it must have taken me fifteen minutes to pay, by which time I was hot and cross.

I had bought a new dress for the party—new to me, anyway. It came from the UFF charity store on Vasagatan, a boat-necked sheath dress that was probably from the sixties, made of a shiny brocade fabric: dark blue with an abstract pattern that resembled cartoon shooting stars in gold and cerise and turquoise, colors that remined me of Andreas's grandmother's baubles. I pictured my hair in a chignon, I would wear my high-heeled shoes, maybe even turquoise eye shadow to match the stars on my dress.

The crowds diminished as I made my way up the street, they eased as soon as I crossed Kungsgatan and when I turned into Kammakargatan, I felt the calmness envelop me. The stars between the buildings shone with their soft yellow light, and outside the door stood Roman, who had come from the other direction. He was laden with shopping, I took one of the Systembolaget bags so that he could key in the door code.

"We're serving German mulled wine—Glühwein—at the party," he said as we walked in. "I can't bear that sickly sweet Swedish *glögg*."

I followed him into the hallway, we took off our outdoor clothes and went into the kitchen where he began to unpack the bag of groceries: several nets of oranges, packets of cinnamon sticks and cloves. The Systembolaget bags contained twelve bottles of German Pinot Noir.

"Twelve bottles should be enough, right?" he said. "Most people will only have one or two cups of Glühwein. There's plenty of other stuff to drink too."

"I don't know how many you're expecting," I said.

He looked thoughtful.

"Maybe forty people?"

"Forty?"

"Or fifty?"

"So how big is this party?"

"You know how it is," he said, putting on an apron. "If we invite thirty, at least forty will show up, and as we've invited forty, I guess there will be at least fifty. Maybe more. Do you want to help me?"

I nodded, he smiled at me and passed me an apron, then he switched on the radio in the window. It was always tuned to P2, and a piece of classical music filled the air.

"Can you take care of the oranges?"

"Sure."

"Rinse them first, then slice them."

I cut open the red net around the oranges with a knife, rinsed them thoroughly then dried them on an old kitchen towel that had been pressed so smooth that it shone. Then I placed one orange on the old wooden chopping board on the counter.

"How thick do you want the slices?"

Roman frowned. "Maybe half an inch," he said hesitantly. "Or quarter. Wait a minute."

He wiped his hands and came to stand beside me, he reached for the knife that I was still holding, and because I hadn't realized that he was going to try to take it I didn't let go, and he placed his hand over mine.

"I was just going to . . . ," he said, at the same time as I think I said "Oh, right," and let go of the knife. We glanced at each other, both smiling, surprised by the unexpected contact.

He cleared his throat, took the knife and began to slice the orange. He held up a slice. "This is what I had in mind," he said quietly.

There were a number of knives in the large block, so I took out another one and began to slice my orange next to his on the chopping board. The smell filled the entire kitchen, it was just on the borderline between sweet and fresh, the juice was sticky on my fingers, it trickled down my wrist and beneath the sleeve of my sweater. Roman reached for another orange, his hand brushed fleetingly against mine again. The strings kicked in on the radio.

At that moment there was a rattle from the front door lock and we heard Andreas step into the hallway. As if on a given signal Roman and I took a step away from each other, then he quickly put down his knife and went to rinse his hands under the faucet.

Andreas stuck his head around the kitchen door.

"Oh hi," he said when he saw me. "What are you two doing?"

"We're preparing the Glühwein," Roman said. "Do you want to help?"

Andreas picked up one of the bottles on the kitchen table and inspected the label.

"I can help out by tasting the wine," he said.

21

ANDREAS AND ROMAN MUST HAVE WOKEN early on December 21. I saw them from my window when I was sitting at the little table having breakfast: two small figures moving back and forth across the snow-covered courtyard, putting empty boxes in the storage area for bulky waste, working together to hang up one of the large living room rugs on the carpet beating rack, whereupon Roman rather clumsily started walloping it with an old rattan beater.

In the afternoon I went down and asked if I could help with anything. Andreas was stressed and barely said a word, Roman was in more of a flap, with an unbridled energy that filled the whole apartment.

"Can you make clove-studded oranges?" he asked. "We've forgotten them."

He sounded upset, so I sat down at the table and pressed sharp-pointed cloves into the oranges that were left over from the Glühwein, then I arranged them in a glass fruit bowl along with some beautiful red winter apples I found in the kitchen, then placed the bowl on the living room table.

This seemed to have a calming effect on him. He was setting out cups and glasses on the kitchen table, while Andreas was busy tidying away shoes and jackets in the hall, I could hear him muttering.

When they were about to vacuum, I went upstairs to get changed. The dress really did fit me very well, it looked as if it had been tailor-made for someone with exactly the same measurements as me back in the sixties. I applied my make-up carefully, tried to put my hair up in a chignon, which isn't easy when you have curly hair, it didn't quite turn out like a chignon, but it looked pretty good anyway. When I went back downstairs to Andreas and Roman a few hours later the apartment was sparkling clean, everything seemed to shimmer and shine. They had trimmed the Christmas tree the previous evening and it was perfect, covered with the prettiest decorations from the boxes, there was the sweet smell of resin, with a hint of smoke from the open fireplace, where Andreas was lighting a fire. Roman was sitting on the sofa, he had changed into a white shirt and a black velvet jacket. On the table was a jug of Glühwein and three cups.

"We'd better taste our creation before everyone arrives," he said, filling one of the cups and handing it to me.

It was surprisingly delicious. Nowhere near as sweet as *glögg*, the Swedish equivalent, not as sour as wine, but fresh, almost effervescent, with the warm, spicy aromas of cinnamon and clove and orange. It felt as if the very essence of Christmas had been captured in that little cup.

"It's really good," I said.

He nodded in agreement. "We did well."

22

THE FIRST GUESTS ARRIVED AT SEVEN THIRTY, the rest trickled in during the evening. The living room quickly filled up with people drinking Glühwein, helping themselves to canapés and Christmas treats, talking and laughing. I found myself in conversation with David about Magnus Dahlström, whom to my surprise he had read and liked, and we agreed that *Fyr* was a masterpiece.

"I don't think I've ever read anything with such a dense atmosphere," I said, and he nodded enthusiastically.

"I was actually heavily influenced by the book when I wrote the lyrics to our second album," he said, but as I hadn't heard Intermission's second album I couldn't judge how successful he had been.

Andreas appeared and topped off our glasses. He was in a good mood, his pre-party stress had melted away as soon as the first guest appeared, and when the doorbell rang again he hurried off to answer it.

I didn't even need to turn my head to know who had arrived, I could tell from his voice as soon as he opened the door: he sounded pleasantly surprised but there was a distance in his tone, and I thought the atmosphere in the entire apartment changed, spreading like a light frost into the warmth of the living room.

It was Louise. She was wearing a short-sleeved white angora sweater, I had never seen a sweater that looked so soft. She had paired it with an ice-blue skirt that shone like silk, skimming her hips then flowing outward like waves around her legs. She looked expensive, expensive and modern, and I suddenly felt out of place in my dress, with my vain attempt to appear elegant when in fact I was grubby and poor in my second-hand clothes.

Roman was standing in the doorway between the kitchen and the living room when she walked in, and as soon as he saw her he stiffened, only for a second, then he pulled himself together and came to meet her with a smile.

"I didn't think you'd come," he said.

"I can't stay long," she replied.

"What would you like to drink? Some Glühwein?"

She shook her head so discreetly that it was no more than a slight twitch, a rapid movement that meant no thanks, no, I would prefer not to drink that kind of thing, and with lightning speed Roman went from wanting to tell her how delicious our creation was to adopting an understanding expression: of course she didn't want Glühwein, what would she like instead?

"Champagne?" he said and she didn't even nod, just gave him a faint smile that meant yes, and they went into the kitchen together.

"Have you read *Hem*?" David asked.

He wanted to continue our conversation about Magnus Dahlström. I nodded.

"It's funny," I said, "I read it at the beginning of the fall when I had just left home for the first time. So the whole thing felt especially relevant."

The comment had sounded clever in my head a second ago, but as soon as I uttered the words I wondered what I actually meant. How drunk was I? I could barely remember what the book was about. Did I even finish it? Most of all, I wondered what Roman and Louise were doing in the kitchen.

"I'm not sure how to interpret the title," I said, hoping to move on from what I had just said. "Is it home as in the direction, homeward, or home as in a home, or several homes, in the plural?"

David looked puzzled.

"I think I'll have another drink," I said. "Would you like one too?"

We both went into the kitchen. There was no sign of Roman and Louise, which was strange, but then I spotted something white out on the balcony.

They were standing there, very close to each other, talking, they seemed to be upset about something. Louise was gesticulating with her lit cigarette, the glowing end darting around like a firefly in the darkness.

"I think it's homes, as in several homes," David said.

I filled my glass from the first bottle at hand, a smooth white wine that had become lukewarm. Suddenly the balcony door flew open and Louise stepped into the kitchen, she looked as if she'd been crying. Roman followed, his face pale. I caught his eye for a second before he turned away and hurried after Louise, who was heading for the hallway. I saw her hunting among the coats for her fur, she pulled it on quickly then vanished through the front door, leaving Roman standing helplessly in the living room watching her go. Andreas got up from the sofa and went over to him.

"Can I talk to you?" he said.

"Maybe we can do this some other time when it's more convenient," Roman replied quietly.

"I'd prefer to talk now," Andreas said.

They disappeared into the area of the apartment where the bedrooms were, I had only been there a few times to pick something up or use the bathroom. The sound of muffled but agitated voices drifted toward me.

Suddenly I felt sick. I rarely drank enough to make me feel that way, but maybe I'd had too many drinks during the course of the evening. I tried to count them but lost track, there was the Glühwein on the sofa before the guests arrived, then a glass of wine, still as a kind of pre-party drink, actually it might have been two glasses, then more Glühwein as people began to show up, at least two cups, maybe three, then more wine, several glasses, quite a lot of glasses I realized as the nausea quickly became worse and I had to leave the party and hurry up the stairs to my room.

23

WHEN I WOKE UP I WAS LYING ON MY BED FULLY dressed. It was just after eleven o'clock in the morning and the room was filled with light, I hadn't drawn the blinds before I fell asleep. I had a terrible hangover.

I pulled the covers over my head and tried to go back to sleep, but the room was too light. I flung open the window, greedily drew the cold air into my lungs and immediately felt my head clear a little. After a while I got up and changed my clothes, removed the worst of the smeared make-up from the previous evening, then went for a quick walk around Tegnérlunden, where the day was in full swing. The sun was shining mercilessly and I hadn't thought to bring my sunglasses, children and dogs were running around chasing one another, there were children sledding, throwing snowballs, laughing, swarming about like in a Bruegel painting.

On the way back I rang Andreas and Roman's bell, but no one answered. That was strange. Were they still asleep? They always used to wake up early after parties, it was as if the alcohol didn't affect them as much as it did me, and by this time they should be busy cleaning and washing dishes and heading for the recycling with the empty wine bottles.

A few hours later, when dusk was falling and my nausea had subsided and been replaced by the need for something salty and greasy, I went out again, to the pizzeria on Dalagatan, where I bought a kebab, a box of French fries and a Coke, they put it all in a bag for me and I headed home. Kammakargatan lay silent and deserted, a lot of places were in darkness, but up in Andreas and Roman's apartment the old Advent candle bridges glowed in the windows. When I passed their door I rang the bell again, but still no one answered.

The following day was the day before Christmas Eve and I woke early, feeling bright and full of energy now that my hangover was gone. On the way out I tried Andreas and Roman again, then I peered through the letterbox but couldn't see anything except the small oriental rug just inside the door.

I went to Nyberg's bakery on Upplandsgatan for breakfast, fresh coffee in a thick bone china cup, a plain cheese roll and a Pago orange juice, it was cheap but felt luxurious and the place was almost empty, a lot of people had already left the city for

Christmas and I was able to sit at the best table by the window. The morning papers lay beside the hotplate for the coffee pots, along with a small silver sparkly Christmas tree, I picked up a copy of *Dagens Nyheter* and flicked idly through the culture section, wishing I had company, someone to sound off to about a facile review of a new French novel. Then I strolled down Tegnérgatan toward Birger Jarlsgatan and on to the Östermalm Food Hall, just as I had planned. The festive atmosphere inside was overwhelming, garlands hung from the ceiling, the counters were overflowing with Christmas food, one even had a whole pig's head on display, glazed and with a red apple in its mouth.

I bought my herring and my salmon and my Jansson's temptation, some meatballs and the obligatory slice of Christmas ham, a small loaf of spiced wort bread and a few handmade Mozart balls and marzipan elves, then I went to the liquor store on Grev Turegatan and bought wine: Champagne, red Burgundy, something white that was more expensive than the type I usually allowed myself.

As I was walking home along Kammakargatan I bumped into David. He looked pleased to see me, gave me a hug and wished me a Merry Christmas.

"Have you heard anything from Andreas or Roman?" I asked. "I don't understand where they've gone."

He shook his head. “I don’t know exactly what happened at the party, but they were arguing about something. I’ve heard them argue before, but it was worse this time. Things actually got pretty weird, I left quite early.”

“But where are they now?”

“They’ve probably had enough of each other and gone home to their parents. It’s happened before. Besides, that’s probably where they’re spending Christmas anyway.”

He didn’t sound particularly concerned.

On the way up to my room I rang their bell again. I opened the letter box and shouted into the apartment, but no one answered.

24

AND SO I WOKE UP ON THE MORNING OF CHRISTmas Eve, and it had snowed. The Christmas party suddenly seemed very distant, as if the snow had washed everything clean and transformed life into a single huge, silently shimmering moment. I stayed in bed for a long time, getting up only to make breakfast, I drank my coffee in bed, with warm milk, and ate a sandwich made from the spiced bread I had bought, and I opened Thomas Mann's *The Magic Mountain* and began to read. *A perfectly ordinary young man set off one summer's day to travel from Hamburg, his hometown, to Davos-Platz in Graubünden. He was intending to stay for three weeks.*

Time disappeared, soon it was early afternoon and I put on a record of Christmas music, opened my bottle of Champagne and slowly sipped a glass while I prepared the food. It was still snowing outside the window, I set the table beautifully then ate my little Christmas lunch, and as I sat there the bells of St. Peter's church began to ring, and everything felt somehow solemn and dignified.

As twilight began to fall I decided to go for a walk. I put on my coat and shoes, and when I opened the door I saw a small package on the floor of the stairwell. It was wrapped in thick brown paper with a dark red bow.

I picked it up, went back inside and sat down on the bed, weighing the package in my hand. A Christmas present. Carefully I untied the bow and removed the paper from what turned out to be a box. Inside, on a bed of moss-green velvet, lay a silver bracelet. It was a chain of links in the shape of elegantly stylized clusters of grapes and leaves, possibly from the 1930s. I had never even held such a beautiful piece of jewelry in my hand before, let alone owned one, and I examined both the box and the wrapping paper in order to try to find out who it was from, but there was no label, no message, no clues.

I looked out of the window, it was still snowing: light, fluffy flakes, the roofs of the buildings were white and the lights on the tall fir tree in the courtyard shone softly beneath a blanket of snow. The whole neighborhood seemed to have settled into a quiet Christmas peace.

I held up the bracelet and gazed at it in the last of the pale daylight. It wasn't the kind of thing I would ever have thought of acquiring for myself, but as I contemplated it now I had a strong feeling that it had come to me for a reason. I almost thought it was whispering to me, telling me that of course I could wear a bracelet like that, of course I could be a person who was entitled to demand what is beautiful in the world. It was cool beneath my fingertips, and cool against my skin when I carefully placed it around my wrist and fastened it.

(And I would be lying if I said that I had never taken it off since that day, because for long periods I didn't wear it at all, but I am looking at it now, on my left wrist, as I write this: a text whose purpose I am not even sure of myself. Call it a mid-life crisis, call it sentimentality—perhaps both are equally true. All I know is that I suddenly felt a powerful, almost irresistible need to write down what I still think I can remember from those months on Kammakargatan. It is a time I carry with me in the same way as that unexpected Christmas present: like a discreet, subtly shining piece of jewelry, like a gift I could never have expected, but which in retrospect I cannot imagine being without. In many ways it was a parenthesis in my life, but if there is one thing I have learned since then, it is that the most significant things in life sometimes take place in parentheses.)

ABOUT THE AUTHOR

THERESE BOHMAN GREW UP OUTSIDE OF NORR-köping and now lives in Stockholm. Her debut novel, *Drowned*, received critical acclaim both in Sweden and internationally, and was selected as an Oprah Winfrey Summer Read. Her second novel, *The Other Woman* (Other Press, 2014), was short-listed for the Nordic Council Prize and Swedish Radio's Fiction Prize, while her third novel, *Eventide* (Other Press, 2016), was short-listed for Sweden's most prestigious literary award, the August Prize. Her other novels include *Andromeda* (Other Press, 2025) and *Mount Verity* (Other Press, 2026). Bohman is an arts journalist who regularly contributes to one of Sweden's largest newspapers, *Expressen*.

ABOUT THE TRANSLATOR

MARLAINE DELARGY STUDIED SWEDISH AND German at the University of Wales, Aberystwyth, and she taught German for almost twenty years. She has translated novels by many authors, including Viveca Sten; Kristina Ohlsson; Helene Tursten; John Ajvide Lindqvist; Theodor Kallifatides; Johan Theorin, with whom she won the Crime Writers' Association International Dagger in 2010; and Henning Mankell, with whom she won the Crime Writers' Association International Dagger in 2018.